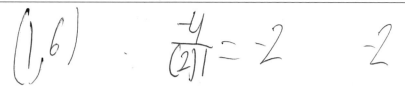

MASTERING ESSENTIAL MATH SKILLS

No-Nonsense Algebra
Practice Workbook

The Perfect Companion to the No-Nonsense Algebra Text

Richard W. Fisher

America's Math Teacher, Richard W. Fisher, will carefully guide you through each and every topic with his award-winning system of teaching.

Edited by Christopher Manhoff

Math Essentials
LOS GATOS, CALIFORNIA

HOW TO GET THE MOST FROM USING THIS BOOK

- Do your work neatly on a separate piece of paper.

- Use graph paper for problems that involve graphing. This will ensure accurate graphs.

- This book can be used alone. For maximum learning, combine this practice workbook with the original *No-Nonsense Algebra* text.

- When you practice, have your original *No-Nonsense Algebra* text open to the lesson that corresponds to the practice lesson in this book. Use the *Introductions, Examples*, and *Video Tutorials* to help you with your practice.

- That's it! This book is designed to be easy-to-use. So get started with this book along with the *No-Nonsense Algebra* text and watch your understanding and progress in Algebra soar!

- Remember that Practice Makes Perfect!

OTHER MATH ESSENTIALS TITLES

Mastering Essential Math Skills: Book 1/Grades 4-5

Mastering Essential Math Skills: Book 2/Middle Grades/High School

Whole Numbers and Integers

Fractions

Decimals and Percents

Geometry

Problem Solving

Pre-Algebra Concepts

No-Nonsense Algebra

The Ultimate Math Survival Guide, Part I/Grades 4-8

The Ultimate Math Survival Guide, Part 2/Grades 4-8

For more information go to www.mathessentials.net

Table of Contents

Chapter 1: Necessary Tools for Algebra

Chapter 2: Solving Equations

Chapter 3: Graphing and Analyzing Linear Equations

Chapter 4: Solving and Graphing Inequalities

Chapter 5: Systems of Linear Equations and Inequalities

Chapter 6: Polynomials

Chapter 7: Rational Expressions (Algebraic Fractions)

Chapter 8: Radical Expressions and Geometry

Chapter 9: Quadratic Equations

Chapter 10: Algebra Word Problems

7/23/18

Add each of the following:

1) -26 + 19

2) -18 + 25

3) 76 + -87

4) -39 + -62

5) -96 + 103

6) -626 + 342

7) -63 + -97

8) -349 + -632

9) -632 + -908

10) -346 + 800

11) -96 + 13 + -12

12) 37 + -18 + 15 + -19

13) -49 + -16 + 92 + -16

14) 363 + -95 + -726

15) -37 + 23 + -16 + 530

1)	-7
2)	7
3)	11
4)	101
5)	7
6)	-284
7)	-160
8)	-981
9)	-1540
10)	454
11)	-95
12)	5
13)	21
14)	-458
15)	500

Score:

1-2 Subtracting Integers

7/23/18

Subtract each of the following.

1) -7 – 9

2) 7 – 12

3) 16 – 19

4) 5 – -6

5) -17 – -26

6) -17 – 18

7) 33 – -16

8) -96 – 72

9) 43 – 16

10) -12 – -18

11) 3 – 7 – -6

12) 72 – 96

13) -99 – 103

14) -76 – 96

15) -363 – -900

16) 43 – -63

17) -42 – -43

18) -302 – 643

19) -4 – 96

20) -96 – -132

1)	~16
2)	~5
3)	-3
4)	17
5)	9
6)	35
7)	49
8)	-168
9)	27
10)	6
11)	2
12)	24
13)	-202
14)	-172
15)	537
16)	106
17)	1
18)	-945
19)	~100
20)	36

Score:

Chapter 1: **NECESSARY TOOLS FOR ALGEBRA**

1-3 Multiplying Integers

Multiply each of the following.

1) 7 x -33

    -231

2) -92 • 3

    -276

3) -8 • -19

    152

4) -23(16)

    -368

5) (-32)(-12)

    384

6) -7 x -63

    441

7) 9(-32)

    -288

8) -24 • 25

    -600

9) (-26)(12)

    -312

10) -6 x -102

    612

11) (-2)(-3)(-6)

    -36

12) 3(-6)(-4)

    72

13) 3(-5) • 4(-6)

    300

14) -4(-3) • 3(-5)

    -180

15) (2)(6)(-2)(-3)

    72

16) 12(-9)(-2)

    432

17) (-1)(-2)(-3)(-4)

    24

18) 7 • (-6)(-4)

    168

19) (-4)(3)(-2)(-1)

    -24

20) (63)(-2)(4)

    504

1)
2)
3)
4)
5)
6)
7)
8)
9)
10)
11)
12)
13)
14)
15)
16)
17)
18)
19)
20)
Score:

1-4 Dividing Integers

7/22/18

Divide each of the following.

1) $\dfrac{-48}{6} = -8$

2) $\dfrac{-80}{-5} = 16$

3) $-72 \div 4$
-18

4) $-384 \div 6 = -16$

5) $-340 \div -2$
170

6) $\dfrac{168}{-12} = -14$

7) $\dfrac{-225}{25} = -9$

8) $\dfrac{368}{-16} = -23$

9) $\dfrac{64 \div -8}{12 \div -3} = 2$

10) $\dfrac{4 \cdot -12}{-10 \div 5} = 24$

11) $\dfrac{6 \cdot (-8)}{2 \cdot (-1)} = 24$

12) $\dfrac{-6 \, (-16)}{-12 \div -6} = 48$

13) $\dfrac{-45 \div -5}{-15 \div -5} = 3$

14) $\dfrac{5 \, (-60)}{-20 \div -2} = -30$

15) $\dfrac{-42 \div 2}{-7 \cdot -3} = -1$

16) $\dfrac{6 \cdot -12}{3 \, (-8)} = 18$

17) $\dfrac{-81 \div -9}{-27 \div 3} = -1$

18) $\dfrac{5 \, (-15)}{20 \div -4} = 15$

19) $\dfrac{-84 \div 2}{-24 \div 4} = 7$

20) $\dfrac{(-6) \, (-3) \, (2)}{(-2) \, (9)} = -2$

1) _____
2) _____
3) _____
4) _____
5) _____
6) _____
7) _____
8) _____
9) _____
10) _____
11) _____
12) _____
13) _____
14) _____
15) _____
16) _____
17) _____
18) _____
19) _____
20) _____

Score: _____

1-5 Positive and Negative Fractions

Simplify each of the following.

1) $-\frac{1}{3} + \frac{1}{2} = 1/6$

2) $\frac{3}{4} + -\frac{2}{5} = 7/20$

3) $-\frac{3}{4} + -\frac{1}{3} = 1\frac{1}{12}$

4) $3\frac{1}{2} + -1\frac{1}{4} = 2\frac{1}{4}$

5) $\frac{1}{3} - \frac{2}{3} = -1/3$

6) $\frac{2}{5} - \frac{3}{4} = 7/20$

7) $-\frac{2}{5} - \frac{1}{2} = 9/10$

8) $\frac{1}{2} + -\frac{1}{3} = 1/6$

9) $\frac{1}{4} - \frac{2}{5} = -3/20$

10) $-\frac{4}{5} + -\frac{1}{3} = -1\frac{2}{15}$

11) $-\frac{3}{4} \times -1\frac{1}{2} = 1\frac{1}{8}$

12) $2\frac{1}{2} \cdot -\frac{1}{2} = -1\frac{1}{4}$

13) $\frac{5}{6} \times -\frac{3}{10} = -1/4$

14) $-2\frac{1}{3} \div -1\frac{1}{7} = 2\frac{1}{4}$

15) $-3 \times 1\frac{1}{2} = -4\frac{1}{2}$

16) $-2\frac{1}{2} \div -1\frac{1}{4} = 2$

17) $-3\frac{1}{2} \div \frac{1}{2} = -7$

18) $-\frac{3}{4} \div -\frac{1}{3} = 2\frac{1}{4}$

19) $\frac{3}{5} + -\frac{1}{2} = 1/10$

20) $-\frac{2}{3} + -\frac{1}{4} = -11/12$

1)	
2)	
3)	
4)	
5)	
6)	
7)	
8)	
9)	
10)	
11)	
12)	
13)	
14)	
15)	
16)	
17)	
18)	
19)	
20)	
Score:	

1-6 Positive and Negative Decimals

Simplify each of the following.

1) -7.26 + 4.9

    -3.36

2) -6.73 + -5.96

    12.69

3) 3.61 + -2.33

    1.28

4) -2.6 + 7.9562

    5.3562

5) -3.9 + -4.37

    8.27

6) 3.1 – 4.63

    -1.53

7) -21.6 – 7.68

    -29.28

8) 3 – 5.67

    -2.67

9) -.7 – .9

    -1.6

10) 3.632 – 7.97

    -4.762

11) 5 x -2.6

    -13

12) -3.36 • -.4

    13.44

13) -2.3 x -4.6

    10.58

14) 1.5 x -2.13

    -3.195

15) 5.32 x -1.2

    -6.384

16) -6.22 ÷ 2

    -3.11

17) -8.112 ÷ .3

    -27.04

18) -5.25 ÷ -5

    1.05

19) -3.148 ÷ .4

    -7.87

20) -6.25 ÷ -.5

    12.5

1)

2)

3)

4)

5)

6)

7)

8)

9)

10)

11)

12)

13)

14)

15)

16)

17)

18)

19)

20)

Score:

1-7 Exponents

For 1-10, rewrite each as an integer.

1) 9^2

81

2) $(-3)^2$

9

3) $(-6)^3$

-216

4) 5^3

25

5) $(-7)^4$

2401

6) 3^4

81

7) $(-8)^3$

-512

8) $(-9)^4$

6561

9) 2^8

256

10) $(-7)^3$

-177

For 11-20, rewrite each as an exponent.

11) 5 x 5 x 5 x 5 x 5

5^5

12) 36

6^2

13) (-2)(-2)(-2)(-2)

-2^4

14) 144

12^2

15) 225

15^2

16) (-9)(-9)(-9)

-9^3

17) 100

10^2

18) 2,500

50^2

19) (-10)(-10)(-10)(-10)

-10^4

20) 49

7^2

1)	
2)	
3)	
4)	
5)	
6)	
7)	
8)	
9)	
10)	
11)	
12)	
13)	
14)	
15)	
16)	
17)	
18)	
19)	
20)	
Score:	

1-8 Laws of Exponents

Simplify each of the following.

1) $6^3 \times 6^5$

2) $(\frac{8}{5})^3$

3) $(4 \times 6)^3$

4) $(5^3)^4$

5) $\frac{3^7}{3^2}$

6) 6^{-4}

7) $7^3 \cdot 7^5$

8) $(3 \cdot 9)^3$

9) $(4^2)^5$

10) 3^{-3}

11) 6^1

12) 12^0

13) $\frac{4^2}{4^4}$

14) $(5^3)^3$

15) 5^{-3}

16) $(\frac{7}{8})^4$

17) $\frac{3^8}{3^2}$

18) $\frac{5^2}{5^5}$

19) $5^3 \cdot 5^2 \cdot 5^4$

20) $\frac{6^5}{6^2}$

1) _____

2) _____

3) _____

4) _____

5) _____

6) _____

7) _____

8) _____

9) _____

10) _____

11) _____

12) _____

13) _____

14) _____

15) _____

16) _____

17) _____

18) _____

19) _____

20) _____

Score: _____

1-9 Square Roots

7/23/18

Simplify each of the following.

1) $\sqrt{36}$

6

2) $\sqrt{121}$

11

3) $\sqrt{2500}$

50

4) $\sqrt{400}$

20

5) $\sqrt{75}$

$5\sqrt{3}$

6) $\sqrt{8}$ $2\sqrt{2}$

7) $\sqrt{\dfrac{144}{9}}$

4

8) $\sqrt{3600}$

60

9) $\sqrt{\dfrac{20}{36}}$

10) $\sqrt{\dfrac{12}{25}}$

11) $\sqrt{50}$ $5\sqrt{2}$

12) $\sqrt{72}$ $6\sqrt{2}$

13) $\sqrt{12}$ $2\sqrt{3}$

14) $\sqrt{200}$ $10\sqrt{2}$

15) $\sqrt{\dfrac{16}{4}}$

2

16) $\sqrt{27}$ $3\sqrt{3}$

17) $\sqrt{125}$ $5\sqrt{5}$

18) $\sqrt{4900}$

70

19) $\sqrt{\dfrac{72}{8}}$

3

20) $\sqrt{\dfrac{25}{36}}$

0.83

1)
2)
3)
4)
5)
6)
7)
8)
9)
10)
11)
12)
13)
14)
15)
16)
17)
18)
19)
20)
Score:

1-10 Order of Operations

Solve each of the following.
Be sure to follow the correct order of operations.

1) $6 + 7 \times 4 - 3$

   31

2) $9 + 2^3 \times 4 - 3$

   38

3) $6^2 (24 \div 6) \div 2$

   72

4) $7(8 + 4) - 6^2$

   36

5) $7 + \{(7 \times 5) + 3\}$

   45

6) $5(6 + -11) + -8$

   -33

7) $5^3 - 6(4 + 3)$

   83

8) $(16 + -7) + 72 \div 3^2$

   17

9) $\dfrac{(16 - 8) + 4^2}{-10 + 3(4 + 2)}$

   3

10) $\dfrac{6^2 + (-9 + 5)}{3(5^2 - 15) - 14}$

   2

11) $7(6 + 7) - 5^2$

   66

12) $5^2(10 + 3) + -8$

   317

13) $15 \div 5 \cdot 4 - 6$

   6

14) $5\{(6 + 2) \cdot 4 - 5\} \div 3$

   45

15) $2(6^2 - 5) - 12$

   50

16) $3 + \{4(8 - 2) \div 6\} - 5$

   2

17) $\dfrac{5^2(4^2)}{2(5 - 1)}$

   50

18) $3^3 + 5(3 - 9)$

   -3

19) $12 + 4 \cdot 6 \div 2 + 5$

   29

20) $9 + 4^2 \times 3 - 12$

   45

1) _____

2) _____

3) _____

4) _____

5) _____

6) _____

7) _____

8) _____

9) _____

10) _____

11) _____

12) _____

13) _____

14) _____

15) _____

16) _____

17) _____

18) _____

19) _____

20) _____

Score: _____

1-11 Properties of Numbers

Name the property that is illustrated.

1) $a \times (b + c) = a \times b + a \times c$

2) $0 + a = a$

3) $a + b = b + a$

4) $a \times b = b \times a$

5) $a + \text{-}a = 0$

6) $a \times \frac{1}{a} = 1 \qquad a \neq 0$

7) $(a + b) + c = a + (b + c)$

8) $1 \times a = a$

9) $19 + \text{-}19 = 0$

10) $0 + (\text{-}12) = \text{-}12$

11) $15 \times 7 = 7 \times 15$

12) $1 \times \frac{3}{4} = \frac{3}{4}$

13) $3(7) + 3(8) = 3(7 + 8)$

14) $62 + 37 = 37 + 62$

15) $3 \times (7 \times 2) = (3 \times 7) \times 2$

16) $c + \text{-}c = 0$

17) $a(b + c) = a(b) + a(c)$

18) $(2 \times 3)\,2 = (3 \times 2)\,2$

19) $m + \text{-}m = 0$

20) $(a + b) + c = (b + a) + c$

1)	
2)	
3)	
4)	
5)	
6)	
7)	
8)	
9)	
10)	
11)	
12)	
13)	
14)	
15)	
16)	
17)	
18)	
19)	
20)	
Score:	

1-12 The Number Line

Use the number line to state the coordinates of the given points.

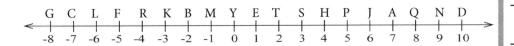

1) B   2) D, E, and K

3) L, S, and P   4) A, P, Q, and M

5) N and A   6) B, C, L, and G

7) A, J, D, and B   8) L, P, H, and A

9) R, P, B, and F

For 10-20, write the numbers in order from least to greatest.
Remember, lower values to the left, and higher values to the right.

10) -15, 20, 5   11) -5, $\frac{1}{5}$, 15

12) $\frac{3}{5}$, $\frac{7}{5}$, 1   13) $\frac{5}{4}$, $\frac{5}{1}$, $\frac{4}{5}$

14) 2^3, 5^2, 3^3 15) $3\frac{1}{3}$, $4\frac{1}{2}$, $2\frac{1}{8}$

16) $\frac{8}{5}$, 1, $\frac{3}{4}$ 17) -103, 100, 180

18) $\frac{1}{2}$, $\frac{3}{4}$, $\frac{1}{4}$ 19) $3\frac{1}{2}$, 2, $1\frac{7}{8}$

20) -7, 9, -15

1)	
2)	
3)	
4)	
5)	
6)	
7)	
8)	
9)	
10)	
11)	
12)	
13)	
14)	
15)	
16)	
17)	
18)	
19)	
20)	
Score:	

Chapter 1: **NECESSARY TOOLS FOR ALGEBRA**

1-13 The Coordinate Plane

Use the coordinate system to find the ordered pair associated with each point.

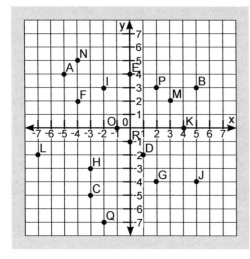

1) A   2) B

3) C   4) D

5) E   6) F

7) G   8) H

9) I   10) J

Use the coordinate system to find the point associated with each ordered pair.

11) (2, 3)   12) (3, 2)

13) (-1, 0)   14) (0, -1)

15) (5, -4)   16) (-4, 5)

17) (-4, 2)   18) (-2, -7)

19) (5, 3)   20) (-7, -2)

1) _____

2) _____

3) _____

4) _____

5) _____

6) _____

7) _____

8) _____

9) _____

10) _____

11) _____

12) _____

13) _____

14) _____

15) _____

16) _____

17) _____

18) _____

19) _____

20) _____

Score: _____

1) Is M = {(2,3), (4,5), (7,8), (7,9)} a function? Why?

2) Is F = {(1,3), (4,7), (3,5)} a function? Why?

3) What is the domain of M?

4) What is the range of M?

5) What is the domain of F?

6) What is the range of F?

7) Explain the vertical line test.

8) All relations are functions. True or False?

9) All functions are relations. True or False?

Use F = {(1,2), (3,5), (4,5), (6,9)} to answer 10-12.

10) List the domain of F.

11) List the range of F.

12) Is F a function? Why?

Use R = {(3,4), (5,6), (5,8), (7,9)} to answer 13-15.

13) List the domain of R.

14) List the range of R.

15) Is R a function? Why?

1) _____

2) _____

3) _____

4) _____

5) _____

6) _____

7) _____

8) _____

9) _____

10) _____

11) _____

12) _____

13) _____

14) _____

15) _____

Score:

_____

1-15 Factors, Divisibility Tests, and Prime Factorization

Use shortcut division to find the prime factorization of each of the following. Express your answers using exponents.

1) 42

2) 36

3) 54

4) 310

5) 125

6) 642

7) 93

8) 98

9) 105

10) 96

11) 144

12) 200

13) 135

14) 216

15) 60

16) 240

17) 127

18) 90

19) 102

20) 180

1) _____

2) _____

3) _____

4) _____

5) _____

6) _____

7) _____

8) _____

9) _____

10) _____

11) _____

12) _____

13) _____

14) _____

15) _____

16) _____

17) _____

18) _____

19) _____

20) _____

Score: _____

1-16 Greatest Common Factors

Find the greatest common factor for each of the following.

1) 10, 15

2) 12, 28

3) 14, 35

4) 18, 24

5) 75, 50

6) 66, 90

7) 70, 108

8) 144, 200

9) 96, 156

10) 22, 33

11) 90, 135

12) 28, 98

13) 15, 24, 36

14) 132, 242

15) 24, 36, 30

16) 105, 210

17) 98, 154

18) 348, 426

19) 105, 126, 210

20) 90, 126, 252

1) _____

2) _____

3) _____

4) _____

5) _____

6) _____

7) _____

8) _____

9) _____

10) _____

11) _____

12) _____

13) _____

14) _____

15) _____

16) _____

17) _____

18) _____

19) _____

20) _____

Score: _____

1-17 Least Common Multiples

Find the least common multiple for each of the following.

1) 12, 15

2) 20, 15

3) 22, 33

4) 14, 70

5) 16, 18

6) 6, 35

7) 24, 20

8) 45, 75

9) 28, 98

10) 15, 22

11) 18, 12

12) 60, 90

13) 5, 6, 10

14) 6, 8, 12

15) 15, 10, 20

16) 30, 24, 40

17) 120, 200

18) 48, 40

19) 100, 56

20) 72, 80

1) _____

2) _____

3) _____

4) _____

5) _____

6) _____

7) _____

8) _____

9) _____

10) _____

11) _____

12) _____

13) _____

14) _____

15) _____

16) _____

17) _____

18) _____

19) _____

20) _____

Score: _____

1-18 Scientific Notation

For 1-10, change each of the following to scientific notation.

1) 14,490,000,000   2) .0000346

3) 259,790   4) .00000721

5) 1,079,000,000   6) .0076

7) .00000019   8) 240,000

9) .0000762   10) 54,000,000

For 11-20, change each number in scientific notation to a conventional number.

11) 6.093×10^5 12) 2.3×10^5

13) 6×10^{-6} 14) 1.347×10^4

15) 3.21×10^5 16) 4.2×10^{-3}

17) 4.5×10^5 18) 5×10^6

19) 3.72×10^{-3} 20) 3.95×10^6

1) _____

2) _____

3) _____

4) _____

5) _____

6) _____

7) _____

8) _____

9) _____

10) _____

11) _____

12) _____

13) _____

14) _____

15) _____

16) _____

17) _____

18) _____

19) _____

20) _____

Score: _____

1-19 Ratios and Proportions

Solve each of the following proportions.

1) $\dfrac{x}{5} = \dfrac{3}{4}$

2) $\dfrac{y}{3} = \dfrac{4}{7}$

3) $\dfrac{5}{x} = \dfrac{3}{2}$

4) $\dfrac{x}{5} = \dfrac{3}{20}$

5) $\dfrac{4}{x} = \dfrac{12}{60}$

6) $\dfrac{16}{x} = \dfrac{12}{9}$

7) $\dfrac{9}{x} = \dfrac{2}{3}$

8) $\dfrac{2}{y} = \dfrac{5}{9}$

9) $\dfrac{8}{5} = \dfrac{10}{x}$

10) $\dfrac{3}{8} = \dfrac{12}{x}$

11) $\dfrac{5}{x} = \dfrac{9}{11}$

12) $\dfrac{2}{7} = \dfrac{5}{x}$

13) $\dfrac{18}{x} = \dfrac{2}{7}$

14) $\dfrac{4}{9} = \dfrac{7}{x}$

15) $\dfrac{100}{x} = \dfrac{90}{45}$

16) $\dfrac{65}{10} = \dfrac{13}{x}$

17) $\dfrac{x}{63} = \dfrac{2}{9}$

18) $\dfrac{4}{5} = \dfrac{28}{x}$

19) $\dfrac{x}{5} = \dfrac{3}{7}$

20) $\dfrac{8}{9} = \dfrac{x}{7}$

1) _____

2) _____

3) _____

4) _____

5) _____

6) _____

7) _____

8) _____

9) _____

10) _____

11) _____

12) _____

13) _____

14) _____

15) _____

16) _____

17) _____

18) _____

19) _____

20) _____

Score: _____

1-20 Using Proportions in Word Problems

Use a proportion to solve each of the following.

1) If 3 tickets to a show cost $66.00, find the cost of 7 tickets.

2) If 3 cans of peas cost $1.65, then how many cans can be bought of $2.75?

3) A man can travel 11 miles on his bike in 2 hours. At this same rate, how far can he travel in 5 hours?

4) A train can travel 90 miles in $1^1/_2$ hours. At this rate, how far will the train travel in 6 hours?

5) A car can travel 150 km on 12 liters of gas. How many liters of gas are needed to travel 500 km?

6) Four cups of sugar are used for every 5 cups of flour in a recipe. How many cups of flour are needed for 64 cups of sugar?

7) How much would you pay for 5 apples if a dozen cost $4.80?

8) A recipe calls for $1^1/_2$ cups of sugar for a 3-pound cake. How many cups of sugar should be used for a 5-pound cake?

9) The scale on a map is 1 inch = 500 miles. If two cities are 875 miles apart, how far apart are they on this map?

10) A man received $420 for working 20 hours. At the same rate of pay, how many hours must he work to earn $735?

11) In a school, the ratio of the number of boys to the number of girls is 5 to 4. If 560 girls attend the school, what is the number of boys attending the school?

12) If 6 apples cost $.99, then how much will ten apples cost?

13) If a watch loses 2 minutes every 15 hours, then how much time will it lose in 2 hours?

14) A picture $3^1/_4$ inches long and $2^1/_8$ inches wide is to be enlarged so that the length will become $6^1/_2$ inches. What will be the width of the enlarged picture?

15) To mix concrete, the ratio of cement to sand is 1 to 4. How many bags of cement would be used with 100 bags of sand?

1) _____

2) _____

3) _____

4) _____

5) _____

6) _____

7) _____

8) _____

9) _____

10) _____

11) _____

12) _____

13) _____

14) _____

15) _____

Score: _____

1-21 Percents

Solve each of the following.

1) Find 20% of 210.

2) Find 6% of 350.

3) 15 is what % of 60?

4) 5 = 20% of what?

5) 15 = 75% of what?

6) 30% of 200 =   .

7) 18 is what % of 24?

8) Find 25% of 64.

9) 3 is what % of 12?

10) 16 is what % of 80?

11) 3 is what % of 60?

12) Find 8% of 320.

13) 20 = 25% of what?

14) Find 40% of 60.

15) 12 = 50% of what?

16) 60 = what % of 80?

17) 3 is 50% of what?

18) Find 100% of 320.

19) 19 is what % of 76?

20) Find 45% of 450.

1) _____

2) _____

3) _____

4) _____

5) _____

6) _____

7) _____

8) _____

9) _____

10) _____

11) _____

12) _____

13) _____

14) _____

15) _____

16) _____

17) _____

18) _____

19) _____

20) _____

Score: _____

1-22 Percent Word Problems

Solve each of the following.

1) Find 20% of 150.

2) 6 is 20% of what?

3) 8 is what % of 40?

4) Change $\frac{18}{20}$ to a percent.

5) A school has 600 students. If 4% are absent, how many students are absent?

6) A quarterback threw 24 passes and 75% were caught. How many were caught?

7) Riley has 250 marbles in his collection. If 50 of them are red, what percent of them are red?

8) A team played 60 games and won 45 of them. What percent did they win?

9) There are 50 sixth graders in a school. This is 20% of the school. How many students are in the school total?

10) A coat is on sale for $20. This is 80% of the regular price. What is the regular price?

11) Steve has finished $\frac{3}{5}$ of his test. What percent of the test has he finished?

12) Alex wants to buy a computer priced at $640. If sales tax is 8%, what is the total cost of the computer?

13) On a test with 120 problems, a student got 90 correct. What percent were correct?

14) A man bought a car for $11,700. If the sales tax was 8%, how much was the sales tax?

15) At a school, 52 students were absent. This was 13% of the students enrolled. How many students are enrolled in the school?

16) A family has a monthly income of $2,500. If 20% of the income is spent on food, how much is spent on food?

17) Alice answered 90% of the 60 problems on a test correctly. How many problems were answered correctly?

18) A road is to be 240 miles long. So far, 180 miles are completed. What percent of the road is completed?

19) A school has an enrollment of 600 students. If 12% of the students are absent, how many are present?

20) There are 40 7th graders in a club. If this is 80% of the club, then how many students are in the club?

1) _____

2) _____

3) _____

4) _____

5) _____

6) _____

7) _____

8) _____

9) _____

10) _____

11) _____

12) _____

13) _____

14) _____

15) _____

16) _____

17) _____

18) _____

19) _____

20) _____

Score: _____

2-1 Solving Equations Using Addition and Subtraction

Solve each equation. Check your answers by substituting them back into the original equation.

1) $n - 3 = 18$

2) $x + 2 = 16$

3) $x - 9 = -7$

4) $x + 15 = -8$

5) $x - 13 = 14$

6) $x - -7 = -8$

7) $-25 = x - 9$

8) $19 = x + 7$

9) $n - 9 = -18$

10) $62 = n - -8$

11) $x + \frac{1}{2} = 14\frac{1}{2}$

12) $x + 2.2 = 7.6$

13) $n - 2.5 = 3.5$

14) $m + 2.2 = -3.9$

15) $2 + x = 3.7$

16) $x - -9 = 2.6$

17) $n + 9 = -13$

18) $n - -9 = -12$

19) $-22 = n - 15$

20) $n - -6.2 = 7.8$

1) _____

2) _____

3) _____

4) _____

5) _____

6) _____

7) _____

8) _____

9) _____

10) _____

11) _____

12) _____

13) _____

14) _____

15) _____

16) _____

17) _____

18) _____

19) _____

20) _____

Score: _____

2-2 Solving Equations Using Multiplication and Division

Solve each equation. Check your answers by substituting them back into the original equation.

1) $7x = 35$

2) $3x = 39$

3) $\frac{x}{7} = 9$

4) $9x = -36$

5) $-12x = 72$

6) $\frac{x}{-3} = 5$

7) $-13x = -104$

8) $\frac{m}{3} = -9$

9) $\frac{m}{-5} = -12$

10) $\frac{-x}{3} = 7$

11) $\frac{-x}{9} = -6$

12) $\frac{1}{3}x = 5$

13) $\frac{2}{3}x = 12$

14) $1\frac{1}{2}x = 12$

15) $6.3x = 44.1$

16) $2.7x = 54$

17) $-\frac{2}{3}x = -30$

18) $-x = 45$

19) $\frac{4}{5}x = 16$

20) $-3.3x = 6.6$

1) _____

2) _____

3) _____

4) _____

5) _____

6) _____

7) _____

8) _____

9) _____

10) _____

11) _____

12) _____

13) _____

14) _____

15) _____

16) _____

17) _____

18) _____

19) _____

20) _____

Score: _____

2-3 Solving 2-Step Equations

Solve each equation. Check your answers by substituting them back into the original equation.

1) $3x + 5 = 35$

2) $4x - 1 = 15$

3) $3x - 5 = 17$

4) $75 = 2x + 11$

5) $\frac{x}{2} + 3 = 9$

6) $\frac{x}{5} - 6 = 5$

7) $\frac{x}{3} + 4 = 13$

8) $\frac{x}{4} - 9 = 51$

9) $12 = \frac{x}{5} + 3$

10) $\frac{x}{2} - 9 = 15$

11) $\frac{x}{7} - 8 = -15$

12) $\frac{2}{3}x - 5 = 5$

13) $3x - 2.5 = -3.7$

14) $2x + 2.8 = 9.6$

15) $4x - -6 = -3.6$

16) $\frac{1}{2}x + 3 = 13$

17) $-2x + 4 = 12$

18) $-3x - 8 = -20$

19) $\frac{2}{5}x + 3 = 11$

20) $-2x - 12 = -20$

1) _____

2) _____

3) _____

4) _____

5) _____

6) _____

7) _____

8) _____

9) _____

10) _____

11) _____

12) _____

13) _____

14) _____

15) _____

16) _____

17) _____

18) _____

19) _____

20) _____

Score: _____

2-4 Solving Equations with Variables on Both Sides

Solve each of the following equations. Check your answers.

1) $5x - 2 = x + 6$

2) $6x + 3 = 2x + 11$

3) $4x - 7 = 3x$

4) $3x + 5 = 4x$

5) $5x + 3 = 2x + 15$

6) $-12x - 65 = -7x$

7) $5x + 10 = 3x + 18$

8) $3x + 90 = 4x + 76$

9) $2x + 24 = -3x + 10$

10) $3x - 12 = 20 - 2x$

11) $51x - 56 = 44x$

12) $9 - 2x = 7x$

13) $23 - 11x = 35x$

14) $8x - 60 = 2x$

15) $5x - 19 = 2x + 12$

16) $9x - 13 = 71 - 5x$

17) $7x + 3 = 2x + 15$

18) $-6x + 40 = 10 - 3x$

19) $9x - 6 = 3x$

20) $2x - 1 = x + 4$

1) _____

2) _____

3) _____

4) _____

5) _____

6) _____

7) _____

8) _____

9) _____

10) _____

11) _____

12) _____

13) _____

14) _____

15) _____

16) _____

17) _____

18) _____

19) _____

20) _____

Score: _____

2-5 Solving Equations Using the Distributive Property

Use the distributive property to solve each of the following.
Check your answers.

1) $5(x + 2) = 20$

2) $3(x - 9) = 30$

3) $7(x + 3) = 28$

4) $4(x - 6) = 44$

5) $5(6x - 2) = 50$

6) $-6(2x + 6) = 48$

7) $-2(3x + 10) = 100$

8) $6(3x - 1) = -42$

9) $7(x + 2) = 5(x + 4)$

10) $3(x - 5) = 2(2x + 1)$

11) $3(x - 5) + 19 = -2$

12) $2(x + 8) - 9 = 5$

13) $7(x - 2) + 17 = 3$

14) $5(2x - 2) + 8 = 43$

15) $3(2x + 1) - 7 = 50$

16) $5(x + 4) = 3(x - 2)$

17) $7(5x - 2) = 6(6x - 1)$

18) $3(2x + 2) = 4(2x - 10)$

19) $5(x + 3) + 9 = 3(x - 2) + 6$

20) $\frac{1}{2}(4x + 4) = \frac{1}{3}(3x + 9)$

1) _____

2) _____

3) _____

4) _____

5) _____

6) _____

7) _____

8) _____

9) _____

10) _____

11) _____

12) _____

13) _____

14) _____

15) _____

16) _____

17) _____

18) _____

19) _____

20) _____

Score: _____

2-6 Solving Equations by Collecting Like Terms

Solve each equation. REMEMBER, if necessary remove parentheses using the distributive property. Check your answers by substituting them back into the original equation.

1) $6x + 3x - x = 80$

2) $5x + 2x - 5 = 30$

3) $9x - 3x - x = 95$

4) $6x + 3x - 4 = 32$

5) $8x - 3x + 3 = 28 - 10$

6) $9x + (2x + 6) = 28$

7) $5x + 2(2x + 4) = 44$

8) $-3x + 6(x + 4) = 9$

9) $3x - 6 - 7x = 12 - x + 6$

10) $5x + 5(1 - x) = x + 8$

11) $3 + 4(x + 2) = 2x + 3(x + 4)$

12) $3(x + 2) - 4x = x + 16$

13) $7x + 4(x + 1) = 15$

14) $6(x + 4) + 2x = 2x + 12 + 2x$

15) $4(x + 2) + 2x = 3(x - 2) + 2x$

16) $32 = 3x + 2x + 7$

17) $6x + 6x + 3 = 5(2x + 1) - 2$

18) $3x - 2 + 4x + 3 = -5 + 3x + 2x + 16$

19) $4(2x + 3) + 2(3x - 4) = 3x - x + 40$

20) $6(x + 3) + 2(2x + 2) = 3(x + 2) + 4(x + 6)$

1) _____

2) _____

3) _____

4) _____

5) _____

6) _____

7) _____

8) _____

9) _____

10) _____

11) _____

12) _____

13) _____

14) _____

15) _____

16) _____

17) _____

18) _____

19) _____

20) _____

Score: _____

2-7 Solving Equations Involving Absolute Value

Solve each equation. Check your answers by substituting them back into the original equation.

1) $|x - 6| = 8$

2) $|x + 3| = 10$

3) $4|x| + 5 = 45$

4) $8|x| + 7 = 71$

5) $|2x - 3| = 13$

6) $|5x + 7| = 32$

7) $|3x| = 15$

8) $|y + 1| = 4$

9) $|2x + 3| = 9$

10) $|5m - 1| = 24$

11) $|4x + 8| = 36$

12) $|5x + 1| = 36$

13) $2|y + 1| = 8$

14) $|3(3x - 5)| = 30$

15) $|x + 4| = 12$

16) $|3x - 10| = 1$

17) $|x - 2| + 3 = 4$

18) $|x + 4| - 1 = 6$

19) $2|3x - 7| + 9 = 17$

20) $|2x + 6| = 10$

1) _____

2) _____

3) _____

4) _____

5) _____

6) _____

7) _____

8) _____

9) _____

10) _____

11) _____

12) _____

13) _____

14) _____

15) _____

16) _____

17) _____

18) _____

19) _____

20) _____

Score: _____

2-8 Simplifying Algebraic Expressions Containing Parentheses

Simplify each of the following.

1) $(7x + 5y) + (-3x - 4y)$ 2) $(9x + 3y) - (4x + 12y)$

3) $(6x - 4y) - (-9x - 7y)$ 4) $(-9x + 12y) + (3x - 15y)$

5) $(3y + 5x - 12) + (4y - 5x - 4)$

6) $(9m - 3n + 5) - (4m + 2n + 6)$

7) $(-6x - 3y + 6) - (9x + 3y + 2)$

8) $(-4x + 3xy + 6) + (-3x + 2y + 2)$

9) $(5x - 3y) - (-2x - 7y)$

10) $(12x + 9y) + (-15x - 30y)$

11) $(-5x - 9y) - (15x + 3y)$

12) $(8x - 2xy - y) + (-7x + 8y)$

13) $(6x - 2y + 2z) - (-10x + 3y - 7z)$

14) $(-7x^2 + 2y - 3x^2y) + (4x^2 - 5y + 2x^2y)$

15) $(6x - 3y - 3) - (-3x - 7y - 9)$

16) $(9x^2 - 7x) - (-6x - 7x^2 + xy)$

17) $(4x^2 - 3xy + x + 7) + (4x^2 + 7x - 9)$

18) $(3x^2 + 2y^2 - 5) - (5x^2 - 5y^2 + 10)$

19) $(9x^2 + 7x - y^2 + 2) - (12x^2 - 5x + 3y^2 - 12)$

20) $(12x^2 - 3xy + 2y^2) + (-3x^2 + 5xy - 7y^2)$

1)	
2)	
3)	
4)	
5)	
6)	
7)	
8)	
9)	
10)	
11)	
12)	
13)	
14)	
15)	
16)	
17)	
18)	
19)	
20)	
Score:	

2-9 Solving Multi-Step Equations

Solve each of the following equations. Check your answers by substituting them back into the original equation.

1) $4(x + 5) = 3(x + 2)$

2) $8 = 4(3x + 5)$

3) $2(x - 3) + 5 = 3(x + 1)$

4) $3(x - 5) + x = x - 6$

5) $2x + 5 = 5(x - 7) - 2x$

6) $\frac{x + 8}{4} = 3$

7) $\frac{2x - 1}{3} = 5$

8) $7 = \frac{9x + 4}{7}$

9) $\frac{x}{2} + 7 = 6$

10) $\frac{2x}{3} - 7 = 13$

11) $7x - 15x - 11 = 9 + 4$

12) $\frac{3x - 2}{2} = 17$

13) $\frac{x}{-4} - 8 = -42$

14) $\frac{x}{3} + 6 = -45$

15) $\frac{4x + 5}{7} = 7$

16) $\frac{7x + (-1)}{8} = 8$

17) $3x + 2(x + 4) = x$

18) $4(x - 3) = 3(x - 1)$

19) $16 + 18 = 8 - 2x$

20) $\frac{x - 5}{4} + 6 = 31$

1) _____

2) _____

3) _____

4) _____

5) _____

6) _____

7) _____

8) _____

9) _____

10) _____

11) _____

12) _____

13) _____

14) _____

15) _____

16) _____

17) _____

18) _____

19) _____

20) _____

Score: _____

3-1 Graphing Linear Equations

Graph each of the following linear equations. Select 4 values for x and find the value for y. Graph and connect the points. Use graph paper.

1) $y = 2x + 2$

2) $y = 3x + 1$

3) $y = 2x - 8$

4) $y = x - 6$

5) $y = \frac{1}{2}x + 4$

6) $y = 2x - 6$

7) $y = -3x + 4$

8) $y = \frac{1}{2}x + 1$

9) $y = -x - 3$

10) $y = -3x + 2$

11) $y = \frac{1}{3}x - 1$

12) $y = 2x + 6$

13) $y = -2x + 4$

14) $y = 2x + 5$

15) $y = x + 6$

16) $y = -3$

17) $y = 2x - 1$

18) $y = \frac{x}{3} + 2$

19) $y = 5x + 3$

20) $y = \frac{1}{2}x + 4$

1) _____

2) _____

3) _____

4) _____

5) _____

6) _____

7) _____

8) _____

9) _____

10) _____

11) _____

12) _____

13) _____

14) _____

15) _____

16) _____

17) _____

18) _____

19) _____

20) _____

Score: _____

3-2 Graphing Linear Equations Using x and y Intercepts

Find the x and y intercepts for each equation and draw the graph. Use graph paper.

1) $x = 5$

2) $y = 4$

3) $x + 3y = 6$

4) $3x - y = 4$

5) $x + y = 2$

6) $x + 5y = 4$

7) $2x - 3y = 6$

8) $5x + 2y = 10$

9) $3x + 2y = 18$

10) $x + 2y = -1$

11) $4x + y = 10$

12) $3x - y = 10$

13) $4x + 5y = 20$

14) $3x - 4y = 12$

15) $2x + 6y = 12$

16) $2x + 3y = 8$

17) $8y + 5x = 24$

18) $2x - 3y = 7$

19) $x + 2y = 8$

20) $2x + y = 6$

1) _____

2) _____

3) _____

4) _____

5) _____

6) _____

7) _____

8) _____

9) _____

10) _____

11) _____

12) _____

13) _____

14) _____

15) _____

16) _____

17) _____

18) _____

19) _____

20) _____

Score: _____

3-3 Slope of a Line

Find the slope of each line that passes through the given points.

1) $(1, 1), (3, 4)$

2) $(8, 4), (6, 5)$

3) $(-4, -1), (-3, -3)$

4) $(-6, 3), (-4, 5)$

5) $(-3, 3), (1, 3)$

6) $(-5, 3), (6, 5)$

7) $(-3, 9), (-7, 6)$

8) $(-2, -3), (0, -1)$

9) $(-2, 1), (-2, 3)$

10) $(4, 8), (1, 3)$

11) $(2, 3), (9, 7)$

12) $(0, -3), (3, -1)$

13) $(2, 6), (-1, 3)$

14) $(5, 7), (-2, -3)$

15) $(4, 0), (5, 7)$

16) $(-8, 3), (-6, 2)$

17) $(3, 2), (5, -6)$

18) $(-2, 0), (1, -1)$

19) $(0, 0), (-3, -9)$

20) $(1, -1), (3, 5)$

1) _____

2) _____

3) _____

4) _____

5) _____

6) _____

7) _____

8) _____

9) _____

10) _____

11) _____

12) _____

13) _____

14) _____

15) _____

16) _____

17) _____

18) _____

19) _____

20) _____

Score: _____

3-4 Changing from Standard Form to the Slope-Intercept Form

Change each of the following linear equations from the standard form to the slope-intercept form and graph the line. Be careful of negative slopes.
Examples: $-\frac{3}{4} = \frac{-3}{4} = \frac{3}{-4}$ The graph will be the same.

1) $y + 2x = 5$

2) $-3x + 2y = 6$

3) $4y + 8x = -16$

4) $6x + 6y = 12$

5) $-3x + y = 6$

6) $-2x + 6y = 18$

7) $4x - 4y = 16$

8) $-4x + 3y = 6$

9) $2y - 3x = 2$

10) $x - 3y = -6$

11) $4x + 2y = 4$

12) $y - 2x = -8$

13) $6x + 3y = 18$

14) $y + 5x = 2$

15) $2x - y = 1$

16) $4x - 3y = 12$

17) $8y + 3x = 24$

18) $y - x = 4$

19) $4y - 2x = 12$

20) $2x + 5y = 10$

1) _____

2) _____

3) _____

4) _____

5) _____

6) _____

7) _____

8) _____

9) _____

10) _____

11) _____

12) _____

13) _____

14) _____

15) _____

16) _____

17) _____

18) _____

19) _____

20) _____

Score: _____

3-5 Determining the Slope-Intercept Equation of a Line

Find the equation of the line in slope-intercept form. Graph the line.

1) Passes through (2, 5) with slope 5.

2) Passes through (5, -2) with slope 3.

3) Passes through (2, 4) with slope $\frac{3}{4}$.

4) Passes through (5, 4) with slope -5.

5) Passes through (2, -6) with slope 1.

6) Passes through (3, 0) with slope -2.

7) Passes through (-3, 0) with slope -3.

8) Passes through (5, 3) with slope $\frac{1}{2}$.

9) Passes through (2, 7) with slope $\frac{5}{6}$.

10) Passes through (3, 5) with slope -2.

11) Passes through (-3, -5) with slope -$\frac{3}{5}$.

12) Passes through (4, -1) with slope $\frac{1}{2}$.

13) Passes through (-$\frac{1}{2}$, 0) with slope 3.

14) Passes through (-3, 3) with slope 1.

15) Passes through (4, -2) with slope 2.

16) Passes through (0, 6) with slope -2.

17) Passes through (3, 7) with slope -3.

18) Passes through (1, 6) with slope $\frac{1}{2}$.

19) Passes through (-3, 5) with slope -1.

20) Passes through (4, -3) with slope -$\frac{3}{5}$.

1) _____

2) _____

3) _____

4) _____

5) _____

6) _____

7) _____

8) _____

9) _____

10) _____

11) _____

12) _____

13) _____

14) _____

15) _____

16) _____

17) _____

18) _____

19) _____

20) _____

Score: _____

3-6 Determining the Point-Slope Form

Find the equation of a line in the point-slope form for each of the following.

$$y - y_1 = m(x - x_1)$$

1) Slope = 5
   passes through (3, 2)

2) Slope = 3
   passes through (4, 5)

3) Slope = 6
   passes through (3, -5)

4) Slope = 4
   passes through (2, 7)

5) Slope = $\frac{3}{5}$
   passes through (-2, 3)

6) Slope = $-\frac{2}{3}$
   passes through (4, -7)

7) Slope = -5
   passes through (-7, 5)

8) Slope = $-\frac{3}{4}$
   passes through (-3, -4)

9) Slope = 5
   passes through (4, 2)

10) Slope = -4
    passes through (4, -6)

11) Passes through
    (3, 5), (6, 7)

12) Passes through
    (-4, 3), (4, -2)

13) Passes through
    (5, 1), (6, -3)

14) Passes through
    (-6, 4), (3, -2)

15) Passes through
    (4, 5), (-2, -3)

16) Passes through
    (1, 2), (4, 6)

17) Slope = 6
    passes through (2, 4)

18) Slope = $-\frac{1}{3}$
    passes through (5, 1)

19) Passes through
    (4, -2), (-4, -5)

20) Passes through
    (-4, -2), (-7, -8)

1) _____

2) _____

3) _____

4) _____

5) _____

6) _____

7) _____

8) _____

9) _____

10) _____

11) _____

12) _____

13) _____

14) _____

15) _____

16) _____

17) _____

18) _____

19) _____

20) _____

Score: _____

4-1 Meaning, Symbols, and Properties of Inequalities

Graph each of the following inequalities.

1) $x < 9$

2) $x > -7$

3) $x \leq 4$

4) $x \geq -9$

5) $x \geq 2.5$

6) $x \leq -3.5$

7) $x > 5\frac{1}{2}$

8) $x \leq -3\frac{1}{2}$

9) $x \geq 0$

10) $x \leq 3.5$

11) $x < 7.5$

12) $x \geq .5$

13) $x \leq 5\frac{1}{2}$

14) $x > -4\frac{1}{2}$

15) $x < 3\frac{1}{4}$

16) $x \geq -3\frac{1}{4}$

17) $x > 10.5$

18) $x \geq -10.5$

19) $x \leq -3$

20) $x \leq 2\frac{1}{2}$

1) _____

2) _____

3) _____

4) _____

5) _____

6) _____

7) _____

8) _____

9) _____

10) _____

11) _____

12) _____

13) _____

14) _____

15) _____

16) _____

17) _____

18) _____

19) _____

20) _____

Score: _____

4-2 Solving Inequalities Using Addition and Subtraction

Solve each inequality. Then graph the solution on a number line.

1) $x + 4 > 12$

2) $x - 3 > -7$

3) $x - 9 \geq -12$

4) $x + 3 \leq 5$

5) $x - 2 < -6$

6) $x + 3 > -2$

7) $x - 3 \leq -7$

8) $x + 6 \geq 16$

9) $x - 12 \leq -8$

10) $x + 7 \geq -7$

11) $x - 1.5 < 4.5$

12) $x - \frac{1}{2} < 6$

13) $x - -9 > 12$

14) $x - -9 \geq 4$

15) $x + \frac{1}{3} > 2\frac{2}{3}$

16) $x - -7 < 9$

17) $x + 2\frac{1}{2} > 3\frac{1}{2}$

18) $x + 1\frac{1}{2} \leq -3\frac{1}{2}$

19) $x - 9 < -8$

20) $x + 2 \leq -4$

1) _____

2) _____

3) _____

4) _____

5) _____

6) _____

7) _____

8) _____

9) _____

10) _____

11) _____

12) _____

13) _____

14) _____

15) _____

16) _____

17) _____

18) _____

19) _____

20) _____

Score: _____

4-3 Solving Inequalities Using Multiplication and Division

Solve each inequality. Then graph the solution on a number line.

1) $3x \geq -24$

2) $-5x \leq -25$

3) $\frac{x}{2} \geq 24$

4) $\frac{x}{-5} < 6$

5) $4x > -12$

6) $-2x < 16$

7) $\frac{x}{-4} < 7$

8) $-12x > 36$

9) $\frac{x}{7} \geq -2$

10) $6x > 42$

11) $-20x < -60$

12) $\frac{1}{3}x < 7$

13) $.5x > 3.5$

14) $\frac{x}{-2} \leq -4$

15) $1.5x \geq 7.5$

16) $-1.5x < 6$

17) $\frac{x}{-2} \geq -3$

18) $\frac{x}{-3} \leq -4$

19) $\frac{x}{3} > -6$

20) $\frac{x}{-7} \geq -2$

1) _____

2) _____

3) _____

4) _____

5) _____

6) _____

7) _____

8) _____

9) _____

10) _____

11) _____

12) _____

13) _____

14) _____

15) _____

16) _____

17) _____

18) _____

19) _____

20) _____

Score: _____

4-4 Solving Multi-Step Inequalities

Solve each inequality. Then graph the solution on a number line.

1) $3x + 4 > 28$

2) $4x + 5 < 37$

3) $-3x + 4 > 13$

4) $\frac{x}{3} - 2 \leq 1$

5) $\frac{x}{-5} - 2 < 3$

6) $-6m + 3 > -9$

7) $5x - 2 \geq x + 8$

8) $-8x + 21 < 6x + 49$

9) $\frac{x}{8} - 13 > -6$

10) $\frac{2x - 3}{5} \geq 7$

11) $6(x - 5) < 35 - 10x$

12) $5x + 21 \geq 3x - 18$

13) $3(x - 7) + 9 \leq 21$

14) $3(x - 4) < 5(x + 4)$

15) $-3x + 6 \leq -3$

16) $7x + 8 > 4(x - 1)$

17) $3x + 19 > -5x + 3$

18) $2(x + 2) \geq 12$

19) $-4x + 7 > 15$

20) $15x - 12 > 7x + 60$

1)
2)
3)
4)
5)
6)
7)
8)
9)
10)
11)
12)
13)
14)
15)
16)
17)
18)
19)
20)
Score:

4-5 Solving Combined Inequalities

Solve each combined inequality. Then draw the graph on a number line.

1) $-1 < x + 3 < 5$

2) $-3 < 2x - 1 < 5$

3) $x - 3 \leq 8$ or $x + 5 \geq 21$

4) $-2 \leq x + 3 < 7$

5) $2x - 3 \leq -6$ or $3x - 12 > 0$

6) $3 \leq x < 12$

7) $3 < x + 2 < 5$

8) $x + 8 < 10$ or $x - 5 > 2$

9) $x + 1 \leq 0$ or $x - 6 > 3$

10) $4x < 16$ or $3x > 15$

11) $-4 \leq x - 3 < 1$

12) $-18 < 3x < 3$

13) $7 < 2x + 5 < 17$

14) $x + 2 < 7$ or $x - 1 > 10$

15) $-5 < 7x - 5 \leq 16$

16) $-6 < 3x - 6 \leq 6$

17) $-1 \leq x + 4 < 4$

18) $-2 < x + 3 < 7$

19) $-10 \leq x - 5 < 8$

20) $x + 2 < 1$ or $x + 3 \geq 5$

1) _____

2) _____

3) _____

4) _____

5) _____

6) _____

7) _____

8) _____

9) _____

10) _____

11) _____

12) _____

13) _____

14) _____

15) _____

16) _____

17) _____

18) _____

19) _____

20) _____

Score: _____

Chapter 4: **SOLVING AND GRAPHING INEQUALITIES**

4-6 Solving Inequalities Involving Absolute Value

Solve each of the inequalities. Then graph the solution on a number line.

1) $|x| > 8$

2) $|x| \le 5$

3) $3|x| < 15$

4) $|x + 5| > 9$

5) $|x - 2| \le 5$

6) $2|x - 3| \ge 8$

7) $|x - 6| \le 6$

8) $|x + 5| < 9$

9) $|2x + 8| \ge 6$

10) $|3x - 8| \ge 5$

11) $|x - 3| < 4$

12) $|x - 3| > 4$

13) $3|x| - 1 < 11$

14) $|2x + 1| \le 5$

15) $|3x| \ge 18$

16) $|2x + 3| \le 9$

17) $|x| < 1$

18) $4|x| \ge 36$

19) $\left|\frac{2x + 1}{3}\right| \ge 5$

20) $|x + 5| \le 3$

1)	
2)	
3)	
4)	
5)	
6)	
7)	
8)	
9)	
10)	
11)	
12)	
13)	
14)	
15)	
16)	
17)	
18)	
19)	
20)	
Score:	

Graph each linear inequality.

1) $y \leq 2x - 1$

2) $y < x + 1$

3) $y > x - 3$

4) $y \leq -3x + 2$

5) $y > -2x + 2$

6) $y \leq x - 2$

7) $y \geq \frac{1}{2}x - 2$

8) $y < -2x + 1$

9) $y \leq 2x - 1$

10) $y > 2$

11) $y > 2x$

12) $y \leq -x + 3$

13) $y \geq \frac{1}{3}x + 2$

14) $y < 2x - 6$

15) $x < 2$

16) $y \leq 2x - 1$

17) $y < -x + 3$

18) $y > 3x + 4$

19) $y > x - 1$

20) $y \leq -2x + 4$

1) _____

2) _____

3) _____

4) _____

5) _____

6) _____

7) _____

8) _____

9) _____

10) _____

11) _____

12) _____

13) _____

14) _____

15) _____

16) _____

17) _____

18) _____

19) _____

20) _____

Score: _____

5-1 The Graphing Method

Use the graphing method to find the ordered pair that is the solution to each system of linear equations. For an accurate graph, use graph paper.

1) $y = 3x$
   $y = -3x + 4$

2) $y - 2x = 8$
   $2x + y = 4$

3) $y = x$
   $y = 2x + 3$

4) $y = x + 4$
   $y = -x$

5) $y = -x + 2$
   $y = -2x + 5$

6) $3x + y = 1$
   $y - x = -3$

7) $y = x - 4$
   $y = -2x + 2$

8) $y = 2x + 7$
   $y = -3x - 3$

9) $y = -x - 3$
   $y = 3x + 5$

10) $y + x = 6$
    $y - 2x = 3$

11) $y = -x - 1$
    $y = 3x + 7$

12) $y = -3x + 2$
    $y = 2x - 3$

13) $x - y = 0$
    $x + y = 4$

14) $x + y = -4$
    $y - 2x = 5$

15) $y - 3x = -1$
    $y + 2x = 4$

16) $y = x + 2$
    $y = 2x$

17) $y + x = 3$
    $y - x = 1$

18) $y = 2x + 5$
    $y = -x - 1$

19) $x + y = -2$
    $y - x = -6$

20) $y - 4x = 0$
    $y + 2x = -6$

1) _____

2) _____

3) _____

4) _____

5) _____

6) _____

7) _____

8) _____

9) _____

10) _____

11) _____

12) _____

13) _____

14) _____

15) _____

16) _____

17) _____

18) _____

19) _____

20) _____

Score: _____

5-2 The Substitution Method

Use the substitution method to solve each system of linear equations.

1) $x + y = 6$
   $x = y + 2$

2) $y = 5x$
   $2x + 3y = 34$

3) $x = y - 1$
   $y = 4 - 2x$

4) $x + y = 4$
   $y = 2x + 1$

5) $y = 3x$
   $x + y = 12$

6) $y = -4x$
   $x - y = -10$

7) $y = 2x + 1$
   $3x + y = 6$

8) $5x - y = 0$
   $x + 2y = 22$

9) $3x - y = 1$
   $2x + 3y = 8$

10) $y + 4x = 5$
    $5x + y = 0$

11) $y = 2x + 1$
    $x + 3y = 10$

12) $y = 3x$
    $y - x = 12$

13) $y = 6x$
    $x + y = 28$

14) $x = y + 2$
    $2y + x = 17$

15) $2x + 3y = 25$
    $3x + 1 = y$

16) $x + y = 5$
    $x = 13 - 3y$

17) $4x + 3y = 38$
    $x + y = 15$

18) $3x + 7y = -6$
    $x - 2y = 11$

19) $x = 3y$
    $x + y = 16$

20) $x - 2y = 7$
    $x = y + 2$

1) _____
2) _____
3) _____
4) _____
5) _____
6) _____
7) _____
8) _____
9) _____
10) _____
11) _____
12) _____
13) _____
14) _____
15) _____
16) _____
17) _____
18) _____
19) _____
20) _____

Score: _____

5-3 The Elimination Method

Solve each of the following using the elimination method.

1) x + y = 10
    x − y = 8

2) x + 4y = 5
    2x − 7y = 9

3) x + y = 8
    -x + 2y = 7

4) x + 3y = 16
    2x − 3y = 4

5) 3x − y = 9
    2x + y = 6

6) 3x − y = 5
    x + 2y = 11

7) 4x + 3y = 7
    -4x + y = 5

8) x − 4y = 3
    -2x + y = 8

9) -x − y = 8
    2x − y = -1

10) 5x + y = 13
    4x − 3y = 18

11) x + 3y = 19
    x − y = -1

12) 5x − 2y = 14
    3x + 4y = -2

13) x + y = 5
    5x − 3y = 17

14) 5x + 4y = 7
    3x + 2y = 3

15) x + 2y = 7
    3x − 2y = 5

16) 2x + y = 5
    5x + 3y = 12

17) 4x − 3y = 9
    2x + 3y = 3

18) 2x + 7y = 17
    3x + 5y = 9

19) -4x + y = 7
    4x + 3y = 5

20) 3x − 2y = 15
    7x − 3y = 15

1)

2)

3)

4)

5)

6)

7)

8)

9)

10)

11)

12)

13)

14)

15)

16)

17)

18)

19)

20)

Score:

5-4 Graphing Systems of Inequalities

Graph each pair of linear inequalities. The solution set of the system is where the shading overlaps. Be sure to use graph paper.

1) $y > 2x - 1$
   $y < -x + 3$

2) $y > 2x + 5$
   $y < -x$

3) $y > x$
   $y > -2x + 4$

4) $y \leq -2x + 4$
   $y \geq 3x + 6$

5) $y > -x - 1$
   $y \leq 2x + 1$

6) $y > 2x$
   $y < -2x + 4$

7) $y \leq 3$
   $y > -x + 2$

8) $y \leq -x + 3$
   $y > x - 2$

9) $y > x$
   $y \leq -2x + 4$

10) $y \leq 2x - 1$
    $y > -x - 2$

1) _____

2) _____

3) _____

4) _____

5) _____

6) _____

7) _____

8) _____

9) _____

10) _____

Score: _____

6-1 Adding Polynomials

Add each of the following.

1) $x^2 + 9x + 6$
   $+ 5x^2 - 3x + 5$
   6x² + 6x + 11

2) $5x^2 - 2x + 3$
   $+ x^2 + 7x - 8$
   6x² + 5x - 5

3) $(2x + 3) + (x + 5)$
   3x + 8

4) $(7x + 4) + (-4x - 2)$
   3x + 2

5) $9x^4 - 3x^3 - 6$
   $+ -4x^4 + 2x^3 - 9$
   5x⁴ - x³ - 15

6) $(-3n^2 + 2n - 8) + (n^2 + 5n + 13)$
   -2n² + 7n + 5

7) $y^2 + 2y + 3$
   $+ y^2 - 8y + 5$
   y⁴ - 6y + 8

8) $(5x^2 - 3x + 12) + (x^2 - 4x + 3)$
   6x² - 7x + 15

9) $(x^2 - 3x + 4) + (3x^2 + 5x - 7)$
   4x² + 2x - 3

10) $32x^2 + 5x - 18$
    $+ -17x^2 + 24x + 26$
    15x² + 29x + 8

11) $(-4x^2 - x - 13) + (5x^2 - 6x + 2)$
    2x² - 8x - 121

12) $x^2 - 6xy + 7y^2$
    $+ 3x^2 + xy - y^2$
    4x² - 5xy + 6y²

13) $13x^2 - 9x + 7$
    $+ 8x^2 - 3x - 12$
    21x² - 12x - 5

14) $(x^2 + 2x + 1) + (3x^2 + 4x - 3)$
    4x² + 6x - 2

15) $-x^2 + 2x - 1$
    $+ x^2 + 8x - 12$
    10x - 13

16) $(x^2 - 2xy + y^2) + (-2x^2 + y^2)$
    -x² - 2xy + 2y²

17) $x^2 - 3xy - y^2$
    $+ 6x^2 + 2xy + 5y^2$
    7x² - 3xy - 4y

18) $-2x^2 + x - 1$
    $+ 4x^2 - x + 5$
    2x² + 4

19) $(2x^2 - 5x - 4) + (3x^2 + x + 7)$
    100x² - 80x - 60

20) $2x + 5y - z$
    $+ 3x + 3y + 2z$
    5x + 8y + z

1)
2)
3)
4)
5)
6)
7)
8)
9)
10)
11)
12)
13)
14)
15)
16)
17)
18)
19)
20)
Score:

6-2 Subtracting Polynomials

Subtract each of the following. Subtract 1-10 vertically.

1) $(3x - 4y) - (x + 5y)$

2) $(x^2 - 2x - 1) - (x^2 + 2x + 1)$

3) $(2x^2 - 5x - 4) - (3x^2 + x + 1)$

4) $(3x^2 + 2x + 1) - (4x^2 - 3x + 2)$

5) $(6x^2 - 5xy + 7y^2) - (3x^2 + 4xy - y^2)$

6) $(8x + 7) - (3x + 2)$

7) $(-5x - 7y + 9) - (2x + 4y - 3)$

8) $(7x^3 - 5y^2 - 2) - (5x^3 - 2y^2 + 4)$

9) $(9x^2y^2 - 5xy + 25y^2) - (5x^2y^2 + 10xy - 9y^2)$

10) $(3x - 4y) - (x + 5y)$

Subtract 11-20 horizontally

11) $(2x - 2) - (x - 4)$

12) $(x^2 + 1) - (2x^2 - 3)$

13) $(2x^2 + 3x - 2) - (-3x^2 + 2x - 5)$

14) $(-3x^2 - 3xy + 2) - (-4x^2 - 2xy - 4)$

15) $(-3x^2 - 4xy - 7) - (x^2 + 2xy + 6)$

16) $(y^2 - 2y - 7) - (9y^2 - 2y + 1)$

17) $(x^3 - 2x^2 - 4) - (-4x^3 + x^2 - 2)$

18) $(-5x^3 - 6) - (-3x^3 + 5)$

19) $(-3x^2 + 2x - 9) - (-3x^2 + 5x - 12)$

20) $(-7x^2 + 2x + 4) - (2x^2 - 1)$

1)	
2)	
3)	
4)	
5)	
6)	
7)	
8)	
9)	
10)	
11)	
12)	
13)	
14)	
15)	
16)	
17)	
18)	
19)	
20)	
Score:	

6-3 Multiplying Monomials

Multiply each of the following.

1) $x^5 \cdot x^7$

2) $x^4 \cdot 5x^3$

3) $(-9x^3)(9x^3)$

4) $(xy)(4y^4)$

5) $(7x^3 y^4)(-7x^5 y)$

6) $(3x^2 y^3)(2x^3 y^2)(3x^2 y^3)$

7) $(3x^2)^3$

8) $(-5x^2)^2$

9) $(x^2 y)(-4x^2 y^3 z^3)$

10) $(5x^3)^3$

11) $(4x^2 y^3)(6x^3 y^2)$

12) $(3x)^2 (3y)^2$

13) $(9x^2)^2$

14) $(-3y)(4xy)(3x^2 y^2)$

15) $(-4x^5 y^3)(5x^5 y^4)$

16) $(-3y^2)(4x^2 y)(2xy^3)$

17) $(-3x^3)^3$

18) $(4x)^2 (3y^3)^2$

19) $(\ x^5)(15x^4)$

20) $(-3xy)(-3xy^2)(-3x^2 y^2)$

1)	
2)	
3)	
4)	
5)	
6)	
7)	
8)	
9)	
10)	
11)	
12)	
13)	
14)	
15)	
16)	
17)	
18)	
19)	
20)	
Score:	

6-4 Dividing Monomials

Divide each of the following. Express your answer with positive exponents.

1) $\dfrac{7x^4}{x^2}$

2) $\dfrac{-27x^3}{3x}$

3) $\dfrac{72x^3y^5}{9xy}$

4) $\dfrac{(x^2y^3)^2}{x^3y^4}$

5) $\dfrac{-15x^3y^3}{x^2y^4}$

6) $\dfrac{70x^2y^3}{7x^3y}$

7) $\dfrac{-32x^2y}{8xy}$

8) $\dfrac{54x^2y^2z^3}{9xyz^4}$

9) $\dfrac{x^2z^3}{x^3yz^4}$

10) $\dfrac{45x^3}{15x}$

11) $\dfrac{(3x^3)^2}{3x^3}$

12) $\dfrac{(4y^3)^2}{(4y^2)^2}$

13) $\dfrac{6x^4y^5}{2x^2y^6}$

14) $\dfrac{-9x^3y^3}{-3x^2y^4}$

15) $\dfrac{8x^5y^{12}}{-2x^3y^{10}}$

16) $\dfrac{18x^3y^4z^7}{-2x^2yz^9}$

17) $\dfrac{-30x^5y}{-5xy^7}$

18) $\dfrac{-12x^8y^3}{4x^6y^4}$

19) $\dfrac{(-2x^2)^2}{x^3}$

20) $\dfrac{35x^{11}y^2}{-7x^7y^9}$

1) _____

2) _____

3) _____

4) _____

5) _____

6) _____

7) _____

8) _____

9) _____

10) _____

11) _____

12) _____

13) _____

14) _____

15) _____

16) _____

17) _____

18) _____

19) _____

20) _____

Score: _____

6-5 Multiplying Polynomials by Monomials

For 1–10 multiply using the vertical method.

1) $(4x - 6)$ and $2x$

2) $(3x + 6)$ and $4x^2$

3) $(-3x^5 - 4x^2)$ and $-5x$

4) $(-3x^2 - 6x + 7)$ and $-5x^2$

5) $(-6x^4 + 3x^2 - 4)$ and $4x^2$

6) $(2y^2 - 4yz + z^3)$ and $3y^2$

7) $7x$ and $(3x^2 + 2xy^2 - 3x^2y)$

8) $2x^2y$ and $(4xy^2 - 2xy + 3x^2y^2)$

9) $10xy$ and $(2xy - 3x^2y + x^2)$

10) $-4x$ and $(5x^2 + 3x^3y + xy)$

For 11–20 multiply using the horizontal method.

11) $(6x^2 + 8y^3)$ and $3y$

12) $(3x^2 - 4x + 9)$ and $-2x^2$

13) $8x^2y$ and $(5x + 2y^2 - 3)$

14) $6x^2$ and $(-11x^2 + 3x + 5)$

15) $-5x^2y$ and $(2y + 5xy - y^2 + x^3)$

16) $5ab(6a^2 - 2b + 3a^2b^2)$

17) -5 and $(-3x^2 - 2xy + 3x^2y^3)$

18) x^3 and $(7xy^2 + 2x^2y^2 - 7)$

19) $2x$ and $(4x^2 + 3x - 5)$

20) $4x^2y^2$ and $(3x^2 + y^2 + 9y^3 + 3x)$

1) _____

2) _____

3) _____

4) _____

5) _____

6) _____

7) _____

8) _____

9) _____

10) _____

11) _____

12) _____

13) _____

14) _____

15) _____

16) _____

17) _____

18) _____

19) _____

20) _____

Score: _____

6-6 Multiplying Binomials

Multiply the following pairs of binomials. As you feel more comfortable, begin to do the steps mentally without showing all the work.

1) (x + 8) (x + 2)

2) (x + 6) (x + 7)

3) (x – 4) (x – 9)

4) (x + 2) (x – 3)

5) (5x – 6) (x + 2)

6) (x – 3) (x – 5)

7) (2x – 5) (x + 7)

8) (x + 12) (3x – 2)

9) (x + 4) (x – 8)

10) (8x + 3) (5x + 2)

11) (8x + 4) (-3x – 2)

12) (6x – 5) (3x – 8)

13) (4x – 3) (x + 4)

14) (2x – 31) (x – 5)

15) (2x + 5y) (2x + 5y)

16) (3x – 7y) (2x + y)

17) (2x – 3) (5x – 9)

18) (x – 9) (x – 9)

19) (x – 3) (7x – 9)

20) (6x – 5y) (2x – 3y)

1) _____

2) _____

3) _____

4) _____

5) _____

6) _____

7) _____

8) _____

9) _____

10) _____

11) _____

12) _____

13) _____

14) _____

15) _____

16) _____

17) _____

18) _____

19) _____

20) _____

Score: _____

6-7 Dividing Polynomials by Monomials

Divide each of the following.

1) $\dfrac{12xy + 27y}{3y}$

2) $\dfrac{x - 2x^2 - x^5}{x}$

3) $\dfrac{15x^3 + 24x^2 - 6x}{3x}$

4) $\dfrac{25x^3 + 15x^2 - 30x}{5x}$

5) $\dfrac{20x^6 - 20x^4 - 5x^2}{-5x^2}$

6) $\dfrac{9x^2y^2 + 3x^2y - 6xy^2}{-3xy}$

7) $\dfrac{-5y^2 + 15y - 25}{-5}$

8) $\dfrac{x^3y^3 - x^2y^2 + xy}{xy}$

9) $\dfrac{15x^4y^4 + 20x^2y^3 - 5x^2y^2}{-5x^2y^2}$

10) $\dfrac{-6x^2y - 12xy^2}{-2xy}$

11) $\dfrac{5x^2y^2 - 15xy^3}{5xy}$

12) $\dfrac{36x^5 + 16x^2y^3 - 12x^2y^2}{-4x^2}$

13) $\dfrac{35x^5 + 40x^3 - 45x}{-5x}$

14) $\dfrac{20x^5 - 28x^4 + 35x^3}{-4x^3}$

15) $\dfrac{9x^2 + 9x - 9x^3}{3x}$

16) $\dfrac{24x^3y^4 - 18x^2y^2 - 6xy}{-6xy}$

17) $\dfrac{4x^2y - 12xy^2}{4xy}$

18) $\dfrac{6x^2y + 8xy^2 + 10xy}{2xy}$

19) $\dfrac{18x^2y^3 - 9x^2y^2 + 27x^3y^2}{9x^2y^2}$

20) $\dfrac{15x^3y^3 - 25x^2y^2 + 20x^3y^3}{5x^2y^2}$

1) _____

2) _____

3) _____

4) _____

5) _____

6) _____

7) _____

8) _____

9) _____

10) _____

11) _____

12) _____

13) _____

14) _____

15) _____

16) _____

17) _____

18) _____

19) _____

20) _____

Score: _____

6-8 Dividing a Polynomial by a Binomial

Find the quotient of each of the following. If there is a remainder, include it in the answer.

1) $(x^2 + 9x + 20) \div (x + 5)$

2) $(x^2 + 6x - 16) \div (x - 2)$

3) $(x^2 - 2x - 35) \div (x + 5)$

4) $\dfrac{(x^2 + 11x + 18)}{x + 9}$

5) $(3x^2 + 14x + 15) \div (x + 3)$

6) $(9x^2 - 16) \div (3x - 4)$

7) $(3x^2 - 8x + 4) \div (3x - 2)$

8) $(21x^2 - 10x + 1) \div (3x - 1)$

9) $(x^2 + 2x - 15) \div (x + 5)$

10) $(15x^2 - 19x - 56) \div (5x + 7)$

11) $(24x^2 + 2x - 15) \div (4x - 3)$

12) $(8x^2 + 6x + 3) \div (4x + 1)$

13) $\dfrac{(15x^2 + 34x - 16)}{5x - 2}$

14) $(n^2 - 3n - 54) \div (n - 9)$

15) $(x^2 - 2x + 5) \div (x + 1)$

16) $(3x^2 + 10x - 9) \div (3x - 2)$

17) $\dfrac{(x^2 - 3x - 9)}{x - 3}$

18) $(8x^2 - 6x - 35) \div (4x + 7)$

19) $(x^2 - 15x - 54) \div (x + 3)$

20) $(x^2 - 7x - 60) \div (x - 12)$

1) _____

2) _____

3) _____

4) _____

5) _____

6) _____

7) _____

8) _____

9) _____

10) _____

11) _____

12) _____

13) _____

14) _____

15) _____

16) _____

17) _____

18) _____

19) _____

20) _____

Score: _____

Chapter 6: **POLYNOMIALS**

6-9 Factoring Using Common Monomial Factors

Factor each of the following.

1) $3x^2 - 3x$

2) $2x^2 + 6x$

3) $28x^2 + 21x^4$

4) $14x^3 - 2x$

5) $12x^2 + 9x + 6$

6) $2x^3y^2 + 6xy^3 + 8xy^2$

7) $16x^6 - 32x^5 - 48x$

8) $5xy^2 + 10x^3y^2 - 20xy^2$

9) $8xy^2 + 24xy$

10) $x + x^2y^2 + x^3y^3$

11) $15x^2y^2 + 25xy + x$

12) $6x^3y^3 + 3x^3y^2 + x^2y^2$

13) $2x^7 - 2x^6 - 16x^5 + 4x^3$

14) $10x^2 + 40x$

15) $15xy - 35xy^2$

16) $25x^2y^2 + 30xy^3$

17) $16x^2 + 8x$

18) $2x^3y^2 - 16xy^2 + 8xy$

19) $27x^2y + 9y^3$

20) $5xy^2 + 10x^2y - 15x^3$

1)	
2)	
3)	
4)	
5)	
6)	
7)	
8)	
9)	
10)	
11)	
12)	
13)	
14)	
15)	
16)	
17)	
18)	
19)	
20)	

Score:

6-10 Factoring Trinomials of the Form $x^2 + bx + c$

Factor each of the following.

1) $x^2 + 8x + 15$

2) $x^2 + 12x + 27$

3) $x^2 + 13x + 30$

4) $x^2 + 6x + 8$

5) $x^2 - 9x + 14$

6) $x^2 + 2x - 15$

7) $x^2 - 5x - 24$

8) $x^2 - 9x - 36$

9) $x^2 + 15x + 36$

10) $x^2 + 2x - 3$

11) $x^2 + 3x - 40$

12) $x^2 - 8x + 7$

13) $x^2 + 12x + 35$

14) $x^2 - 19x + 60$

15) $x^2 - 7x + 10$

16) $x^2 + 6x - 7$

17) $x^2 - x - 42$

18) $x^2 + 6x - 72$

19) $x^2 + 7x + 12$

20) $x^2 + 3x - 54$

1) _____

2) _____

3) _____

4) _____

5) _____

6) _____

7) _____

8) _____

9) _____

10) _____

11) _____

12) _____

13) _____

14) _____

15) _____

16) _____

17) _____

18) _____

19) _____

20) _____

Score: _____

6-11 Factoring the Difference of Two Squares

Factor each of the following.

1) $x^2 - 36$

$(x+6)(x-6)$

2) $100 - x^2$

3) $16n^2 - 9$

$(4n+3)(4n-3)$

4) $4x^2 - 25$

5) $25m^2 - 49$

$(5m+7)(5m-7)$

6) $100x^2 - 25$

7) $x^2 - 9$

$(x+3)(x-3)$

8) $144x^2 - 49y^2$

9) $100x^2 - y^2$

$(10x-y)(10x-y)$

10) $25 - 4y^2$

11) $1 - 9x^2$

$(1+3x)(1-3x)$

12) $49 - x^2y^2$

13) $25m^2n^2 - 16x^2y^2$

14) $16m^2 - 9n^2$

15) $16y^2 - 25$

16) $81y^6 - 25y^2$

17) $x^4 - 16$

18) $64x^2 - 25y^2$

19) $81x^4 - 1$

20) $x^8 - y^6$

1)
2)
3)
4)
5)
6)
7)
8)
9)
10)
11)
12)
13)
14)
15)
16)
17)
18)
19)
20)
Score:

6-12 Factoring Completely

Factor each of the following completely.

1) $2a^2 - 2b^2$

$2(a^2 - b^2)$

2) $3x^2 - 27$

$3(x^2 - 9)$

3) $3x^2 - 6x - 24$

$3(x^2 - 2x - 8)$

4) $x^2 - 16$

$1(x^2 - 16)$

5) $2x^2 - 16x + 30$

$2(x^2 - 8x + 15)$

6) $5x^2 - 20x - 60$

$5(x^2 - 4x - 12)$

7) $2x^2 - 16x + 32$

$2(x^2 - 8x + 16)$

8) $x^3 - 4x$

$x(x^2 - 4)$

9) $6x^2 - 24y^2$

$6(x^2 - 4y^2)$

10) $2x^2 - 162$

$2(x^2 - 81)$

11) $3x^2 + 6x + 3$

$3(x^2 + 2x + 1)$

12) $4x^2 - 6x - 4$

$2(2x^2 - 3x - 2)$

13) $4x^3 - 49x$

$x(4x^2 - 49)$

14) $16x^2 - x^2 y^4$

$x^2(16 - y^4)$

15) $10x^3 - 40x$

$10x(x^2 - 4)$

16) $t^4 - 16$

$1(t^4 - 16)$

17) $3x^4 - 3$

$3(x^4 - 1)$

18) $2x^2 - 128$

$2(x^2 - 64)$

19) $3x^2 - 75$

$3(x^2 - 15)$

20) $7x^2 - 7$

$7(x^2 - 1)$

1) _____

2) _____

3) _____

4) _____

5) _____

6) _____

7) _____

8) _____

9) _____

10) _____

11) _____

12) _____

13) _____

14) _____

15) _____

16) _____

17) _____

18) _____

19) _____

20) _____

Score: _____

6-13 Factoring More Difficult Trinomials

Factor each of the following.

1) $3x^2 + 10x + 3$

$(3x + 5)$

2) $2x^2 + x - 3$

3) $3x^2 - 5x - 2$

4) $3x^2 - 2x - 5$

5) $4x^2 - 12x + 5$

6) $6x^2 + 5x - 4$

7) $2x^2 - 7x - 4$

8) $5x^2 + x - 18$

9) $3x^2 + 4x + 1$

10) $2x^2 - x - 1$

11) $3x^2 - 5x - 2$

12) $3x^2 + 7x + 2$

13) $3x^2 - 8x + 5$

14) $5x^2 + 4x - 1$

15) $7x^2 + 22x + 3$

16) $2x^2 + 7x + 5$

17) $9x^2 + 18x + 8$

18) $3x^2 + 5x + 2$

19) $2x^2 - 9x - 18$

20) $2x^2 + 5x + 2$

1) _____

2) _____

3) _____

4) _____

5) _____

6) _____

7) _____

8) _____

9) _____

10) _____

11) _____

12) _____

13) _____

14) _____

15) _____

16) _____

17) _____

18) _____

19) _____

20) _____

Score: _____

7-1 Simplifying Algebraic Fractions

Simplify each of the following. Express you answer with positive exponents.

1) $\dfrac{36n^6}{6n^4}$

2) $\dfrac{48xy^2}{8y^3}$

3) $\dfrac{50x^3y^2}{2x^2y}$

4) $\dfrac{5x^5y^3}{xyz}$

5) $\dfrac{64x^6y^4}{4x^4y^3}$

6) $\dfrac{-42x^6y^3}{-7x^4y^4}$

7) $\dfrac{-20x^6y^7}{5xy^6}$

8) $\dfrac{3x^2y^3z}{xy^4z^2}$

9) $\dfrac{15x^2y}{3x^3y^2}$

10) $\dfrac{21x^2y^2z^3}{3xy^3z}$

11) $\dfrac{-40xyz}{-5x^2y^3z}$

12) $\dfrac{x^2yz^2}{x^4y^2z^3}$

13) $\dfrac{16xy^4}{4x^3y}$

14) $\dfrac{-45x^2y^3}{-9x^3y^6}$

15) $\dfrac{x^2yz^3}{xy^2z^3}$

16) $\dfrac{24xyz}{8x^2yz^3}$

17) $\dfrac{40xy^2z^2}{10y^3z^3}$

18) $\dfrac{-81x^4y^3z^4}{-3xy^4z^5}$

19) $\dfrac{-35x^2y^2z}{-7xyz^2}$

20) $\dfrac{15x^2y}{3x^3y}$

1) _____

2) _____

3) _____

4) _____

5) _____

6) _____

7) _____

8) _____

9) _____

10) _____

11) _____

12) _____

13) _____

14) _____

15) _____

16) _____

17) _____

18) _____

19) _____

20) _____

Score: _____

7-2 Simplifying Algebraic Fractions with Several Terms

Simplify each of the following.

1) $\dfrac{6(x + 2y)}{9x + 18y}$

2) $\dfrac{3x - 3}{x^2 - 1}$

3) $\dfrac{9x - 18}{3x - 12}$

4) $\dfrac{6y^2 - 6}{27y + 27}$

5) $\dfrac{x^2 + x - 20}{x + 5}$

6) $\dfrac{x^2 + 10x + 16}{x + 2}$

7) $\dfrac{4x + 8}{x^2 + 6x + 8}$

8) $\dfrac{2x - 4}{x^2 + 3x - 10}$

9) $\dfrac{x^2 - 36}{x^2 - 5x - 6}$

10) $\dfrac{x^2 - 9}{x^2 + 6x - 27}$

11) $\dfrac{x^2 + x - 2}{x^2 - 3x + 2}$

12) $\dfrac{x^2 - 2x - 15}{x^2 - x - 12}$

13) $\dfrac{3x - 15}{x^2 - 7x + 10}$

14) $\dfrac{3x^2 + x}{3x + 1}$

15) $\dfrac{x^2 - 2x}{x - 3}$

16) $\dfrac{x^2 - 25}{x^2 + 3x - 10}$

17) $\dfrac{6x + 12}{x^2 - x - 6}$

18) $\dfrac{(x - 3)^2}{x^2 - 9}$

19) $\dfrac{x^2 - 9}{x^2 + 5x + 6}$

20) $\dfrac{3x - 3}{x^2 - 2x + 1}$

1) _____

2) _____

3) _____

4) _____

5) _____

6) _____

7) _____

8) _____

9) _____

10) _____

11) _____

12) _____

13) _____

14) _____

15) _____

16) _____

17) _____

18) _____

19) _____

20) _____

Score: _____

7-3 Using the -1 Factor to Simplify Algebraic Fractions

Simplify each of the following using the -1 factor.

1) $\dfrac{x - y}{y - x}$

2) $\dfrac{x - 3}{3 - x}$

3) $\dfrac{x - y}{2y - 2x}$

4) $\dfrac{x^2 - 25}{5 - x}$

5) $\dfrac{5x^2 - 5}{5 - 5x}$

6) $\dfrac{4 - x}{x^2 - 16}$

7) $\dfrac{2 - x}{x^2 - 4}$

8) $\dfrac{r - s}{s - r}$

9) $\dfrac{2x^2 - 50}{10 - 2x}$

10) $\dfrac{7 - x}{x^2 - 49}$

11) $\dfrac{m^2 - n^2}{n - m}$

12) $\dfrac{x - y}{y - x}$

13) $\dfrac{10x - 10y}{10y - 10x}$

14) $\dfrac{m^2 - 4}{2 - m}$

15) $\dfrac{x^2 - y^2}{5y - 5x}$

16) $\dfrac{6 - x}{x^2 - 36}$

17) $\dfrac{7 - x}{x - 7}$

18) $\dfrac{4x + 2y}{-y - x}$

19) $\dfrac{x - 12}{12 - x}$

20) $\dfrac{x - y}{3y - 3x}$

1) _____

2) _____

3) _____

4) _____

5) _____

6) _____

7) _____

8) _____

9) _____

10) _____

11) _____

12) _____

13) _____

14) _____

15) _____

16) _____

17) _____

18) _____

19) _____

20) _____

Score: _____

7-4 Solving Proportions Containing Algebraic Fractions

Solve each of the following proportions.

1) $\dfrac{6}{x} = \dfrac{2}{9}$

2) $\dfrac{x}{12} = \dfrac{5}{4}$

3) $\dfrac{x}{4} = \dfrac{x+2}{5}$

4) $\dfrac{x-2}{x+1} = \dfrac{1}{2}$

5) $\dfrac{30}{4x} = \dfrac{10}{24}$

6) $\dfrac{5}{x+2} = \dfrac{4}{x}$

7) $\dfrac{x+3}{5} = \dfrac{x+1}{4}$

8) $\dfrac{x}{x+4} = \dfrac{x+1}{x+6}$

9) $\dfrac{x+2}{x-2} = \dfrac{5}{3}$

10) $\dfrac{x+4}{6} = \dfrac{x+6}{4}$

11) $\dfrac{3y-4}{1} = \dfrac{2y}{2}$

12) $\dfrac{5m+2}{3} = \dfrac{3m-1}{2}$

13) $\dfrac{3x+3}{3} = \dfrac{7x-1}{5}$

14) $\dfrac{3x}{2} = \dfrac{15}{5}$

15) $\dfrac{2x}{5} = \dfrac{x+1}{3}$

16) $2 = \dfrac{4x+6}{7}$

17) $\dfrac{5x-3}{4} = \dfrac{5x+3}{6}$

18) $\dfrac{x-5}{4} = \dfrac{3}{2}$

19) $\dfrac{x+1}{x-2} = \dfrac{x+3}{x-1}$

20) $\dfrac{x}{x+4} = \dfrac{x+1}{x+6}$

1) _____

2) _____

3) _____

4) _____

5) _____

6) _____

7) _____

8) _____

9) _____

10) _____

11) _____

12) _____

13) _____

14) _____

15) _____

16) _____

17) _____

18) _____

19) _____

20) _____

Score: _____

7-5 Multiplying Algebraic Fractions

Simplify each of the following.

1) $\dfrac{x}{4} \cdot \dfrac{12}{x}$

2) $\dfrac{32}{x^2} \cdot \dfrac{3x^2}{8}$

3) $\dfrac{8}{x^2} \cdot \dfrac{x^4}{4x}$

4) $\dfrac{10r^3}{6n^3} \cdot \dfrac{42n^2}{35r^2}$

5) $\dfrac{5ab^3}{8c^2} \cdot \dfrac{16c^2}{15a^2b}$

6) $\dfrac{2}{r-2} \cdot \dfrac{r^2-2r}{4}$

7) $\dfrac{12a}{5a} \cdot \dfrac{a^2}{6a^2}$

8) $\dfrac{5}{4x+16} \cdot \dfrac{4}{25}$

9) $\dfrac{x^2-25}{9} \cdot \dfrac{x+5}{x-5}$

10) $\dfrac{y^2-4}{y^2-1} \cdot \dfrac{y+1}{y+2}$

11) $\dfrac{x-5}{x} \cdot \dfrac{x^2}{x^2-2x-15}$

12) $\dfrac{4y}{5} \cdot \dfrac{y-3}{2y}$

13) $\dfrac{5x^2y}{8ab} \cdot \dfrac{12ab}{25x}$

14) $\dfrac{64y^2}{5y} \cdot \dfrac{5y}{8y}$

15) $\dfrac{x-5}{x^2-7x+10} \cdot \dfrac{x^2+x-6}{5}$

16) $\dfrac{x^2-4}{2} \cdot \dfrac{4}{x-2}$

17) $\dfrac{n^2-3n}{n-3} \cdot \dfrac{x^2}{x^2}$

18) $\dfrac{x^2-y^2}{10} \cdot \dfrac{5}{x-y}$

19) $\dfrac{y+6}{2y} \cdot \dfrac{4y^2}{y+6}$

20) $\dfrac{(n-1)(n+1)}{(n+1)} \cdot \dfrac{(n-4)}{(n-1)(n+4)}$

1) _____

2) _____

3) _____

4) _____

5) _____

6) _____

7) _____

8) _____

9) _____

10) _____

11) _____

12) _____

13) _____

14) _____

15) _____

16) _____

17) _____

18) _____

19) _____

20) _____

Score: _____

7-6 Dividing Algebraic Fractions

Divide each of the following.

1) $\dfrac{x}{9} \div \dfrac{x}{3}$

2) $\dfrac{3x}{5y} \div \dfrac{21x}{20y}$

3) $\dfrac{8x^2}{3y^2} \div \dfrac{4x}{6y^3}$

4) $\dfrac{9}{x^2 - 1} \div \dfrac{3}{x + 1}$

5) $\dfrac{6a^2b^2}{8c} \div 3ab$

6) $\dfrac{x^2 - 25}{18} \div \dfrac{x - 5}{27}$

7) $\dfrac{6m^4}{5} \div 2m$

8) $\dfrac{5x - 5}{16} \div \dfrac{x - 1}{6}$

9) $\dfrac{a + 2}{a - 1} \div \dfrac{3a + 6}{a - 5}$

10) $\dfrac{b^2 - 9}{4b} \div (b - 3)$

11) $\dfrac{3x^2}{4y} \div \dfrac{xy}{18}$

12) $\dfrac{x^2 - 1}{2} \div \dfrac{x + 1}{16}$

13) $\dfrac{2x + 2y}{x^2} \div \dfrac{x^2 - y^2}{4x}$

14) $\dfrac{x}{2y} \div \dfrac{xy}{4}$

15) $\dfrac{18}{x^2 - 25} \div \dfrac{24}{x + 5}$

16) $\dfrac{y^2 - 16}{y^2 - 64} \div \dfrac{y + 4}{y - 8}$

17) $\dfrac{x^3}{2y} \div \dfrac{x^2}{4y}$

18) $\dfrac{x^2}{x + 4} \div \dfrac{3x}{x^2 - 16}$

19) $\dfrac{16c^3}{21d^2} \div \dfrac{24c^4}{14d^3}$

20) $\dfrac{8x^2}{x^2 - 25} \div \dfrac{4x}{3x + 15}$

1) _____

2) _____

3) _____

4) _____

5) _____

6) _____

7) _____

8) _____

9) _____

10) _____

11) _____

12) _____

13) _____

14) _____

15) _____

16) _____

17) _____

18) _____

19) _____

20) _____

Score: _____

7-7 Adding and Subtracting Algebraic Fractions with Like Denominators

Add or subtract each of the following.

1) $\dfrac{x}{3} + \dfrac{2x}{3}$

2) $\dfrac{3x}{7} - \dfrac{x}{7}$

3) $\dfrac{5}{2x} + \dfrac{6}{2x}$

4) $\dfrac{3x}{7} - \dfrac{7-2x}{7}$

5) $\dfrac{2x}{x+1} + \dfrac{2}{x+1}$

6) $\dfrac{x}{x-4} - \dfrac{4}{x-4}$

7) $\dfrac{x-2}{x+3} + \dfrac{-1}{x+3}$

8) $\dfrac{4x-5}{x-2} + \dfrac{x+3}{x-2}$

9) $\dfrac{5x-4}{3} - \dfrac{2x+1}{3}$

10) $\dfrac{y^2+5}{y+2} - \dfrac{4y+17}{y+2}$

11) $\dfrac{3a+5}{a-1} + \dfrac{2a-6}{a-1}$

12) $\dfrac{x+5}{6} - \dfrac{x+3}{6}$

13) $\dfrac{b-15}{2b+12} - \dfrac{-3b+8}{2b+12}$

14) $\dfrac{10a-12}{2a-6} - \dfrac{6a}{2a-6}$

15) $\dfrac{2x+3}{x-4} + \dfrac{x-2}{x-4}$

16) $\dfrac{3a+2b}{4ab} + \dfrac{a+2b}{4ab}$

17) $\dfrac{4x+3}{x+2} + \dfrac{3x+4}{x+2}$

18) $\dfrac{y-3}{y+5} - \dfrac{2y-7}{y+5}$

19) $\dfrac{x}{x^2-9} - \dfrac{3}{x^2-9}$

20) $\dfrac{x^2}{x-y} - \dfrac{y^2}{x-y}$

1) _____

2) _____

3) _____

4) _____

5) _____

6) _____

7) _____

8) _____

9) _____

10) _____

11) _____

12) _____

13) _____

14) _____

15) _____

16) _____

17) _____

18) _____

19) _____

20) _____

Score: _____

Chapter 7: **RATIONAL EXPRESSIONS**

7-8 Adding and Subtracting Algebraic Fractions with Unlike Denominators

Add or subtract each of the following. Reduce answers to lowest terms.

1) $\dfrac{2x}{3} + \dfrac{x}{2}$

2) $\dfrac{4x}{3} - \dfrac{x}{6}$

3) $\dfrac{3}{2x} + \dfrac{1}{2}$

4) $\dfrac{1}{5} - \dfrac{2}{y}$

5) $\dfrac{2}{ab} + \dfrac{5}{2a}$

6) $\dfrac{x^2}{2} + \dfrac{3x^2}{8}$

7) $\dfrac{2}{3x} + \dfrac{1}{6xy}$

8) $\dfrac{3x}{8} - \dfrac{x}{12}$

9) $\dfrac{5}{3x} + \dfrac{9}{4x}$

10) $\dfrac{x}{3} - \dfrac{3x}{7} + \dfrac{x}{21}$

11) $\dfrac{4}{5x} + \dfrac{2}{3x}$

12) $\dfrac{5m-1}{6} - \dfrac{3m+2}{9}$

13) $\dfrac{5}{4x} - \dfrac{2}{3x}$

14) $\dfrac{1}{4x} + \dfrac{1}{2x} - \dfrac{2}{3x}$

15) $\dfrac{3}{2x} - \dfrac{1}{8x^2}$

16) $\dfrac{3y-4}{5} - \dfrac{y-2}{4}$

17) $\dfrac{3x-2}{7} + \dfrac{x+1}{14}$

18) $\dfrac{5}{y^2-9} + \dfrac{3}{(y-3)}$

19) $\dfrac{x+y}{2} - \dfrac{2x-y}{6}$

20) $\dfrac{x+y}{8} - \dfrac{x+y}{10}$

1) _____

2) _____

3) _____

4) _____

5) _____

6) _____

7) _____

8) _____

9) _____

10) _____

11) _____

12) _____

13) _____

14) _____

15) _____

16) _____

17) _____

18) _____

19) _____

20) _____

Score: _____

Add or subtract each of the following.

1) $\dfrac{2}{x-1} + \dfrac{2}{x+1}$

2) $\dfrac{3}{x+5} + \dfrac{4}{x-4}$

3) $\dfrac{5}{3x-9} + \dfrac{3}{x-3}$

4) $\dfrac{5}{x-3} - \dfrac{4}{x+1}$

5) $\dfrac{6x}{x-3} + \dfrac{x}{x+1}$

6) $\dfrac{x}{x+5} - \dfrac{3}{x-3}$

7) $\dfrac{2}{x+5} + \dfrac{3}{4x}$

8) $\dfrac{x+1}{x} - \dfrac{x}{x+1}$

9) $\dfrac{4x}{x^2-25} + \dfrac{x}{x+5}$

10) $\dfrac{x}{x^2-1} + \dfrac{4}{x+1}$

11) $\dfrac{4}{x} + \dfrac{3}{x+2}$

12) $\dfrac{x}{x^2-4} - \dfrac{4}{x+2}$

13) $\dfrac{2}{x-4} + \dfrac{3}{x+2}$

14) $\dfrac{2x}{2x+5} - \dfrac{5}{2x+5}$

15) $\dfrac{x}{x+6} + \dfrac{2x}{x-1}$

16) $\dfrac{8x}{x^2-16} - \dfrac{5}{x+4}$

17) $\dfrac{8}{(x+y)^2} + \dfrac{5}{(x+y)}$

18) $\dfrac{4x}{x+1} - \dfrac{x}{x-1}$

19) $\dfrac{1}{x^2-y^2} + \dfrac{3}{(x-y)}$

20) $\dfrac{5}{x} + \dfrac{3}{x+1}$

1) _____

2) _____

3) _____

4) _____

5) _____

6) _____

7) _____

8) _____

9) _____

10) _____

11) _____

12) _____

13) _____

14) _____

15) _____

16) _____

17) _____

18) _____

19) _____

20) _____

Score: _____

7-10 Solving Equations Involving Algebraic Fractions

Solve each of the following.

1) $\frac{x}{3} - \frac{x}{5} = 4$

2) $\frac{x}{2} + \frac{x}{3} = 10$

3) $\frac{2}{5} + \frac{3}{10} = \frac{7}{x}$

4) $\frac{x}{4} + 5 = \frac{3x}{2}$

5) $\frac{x}{6} + \frac{x}{3} = 6$

6) $\frac{x-1}{4} - \frac{2x-3}{4} = 5$

7) $\frac{x+3}{8} - \frac{x-2}{6} = 1$

8) $\frac{x}{x+2} = \frac{3}{7}$

9) $\frac{2x}{3} = 24$

10) $\frac{x}{3} + \frac{x}{7} = 10$

11) $\frac{6x-3}{2} - \frac{x+2}{5} = \frac{37}{10}$

12) $\frac{x-3}{6} - \frac{x-25}{5} = 4$

13) $\frac{x+2}{4} = \frac{5}{2}$

14) $\frac{3x+1}{4} = \frac{44-x}{5}$

15) $\frac{3x-1}{5} = 7$

16) $\frac{2x}{3} - \frac{x}{4} = 10$

17) $\frac{4x-3}{x} = \frac{17}{x}$

18) $\frac{x}{3} + \frac{x}{4} = \frac{7}{4}$

19) $\frac{3y}{8} + \frac{y}{4} = 10$

20) $\frac{1}{6} + \frac{1}{2} = \frac{x}{3}$

1) _____

2) _____

3) _____

4) _____

5) _____

6) _____

7) _____

8) _____

9) _____

10) _____

11) _____

12) _____

13) _____

14) _____

15) _____

16) _____

17) _____

18) _____

19) _____

20) _____

Score: _____

8-1 Simplifying Radical Expressions

Simplify each radical expression.

1) $\sqrt{36}$

2) $\sqrt{144}$

3) $\sqrt{256}$

4) $\sqrt{81x^2}$

5) $\sqrt{18}$

6) $\sqrt{24}$

7) $\sqrt{49x^2 y^2}$

8) $\sqrt{4x^2 y^3}$

9) $\sqrt{48x}$

10) $\sqrt{144x^3 y^2}$

11) $\sqrt{8x^2}$

12) $\sqrt{x^6}$

13) $\sqrt{72x^2}$

14) $\sqrt{36x^3}$

15) $\sqrt{x^3 y^3}$

16) $\sqrt{36x^2 y}$

17) $\sqrt{12x^8}$

18) $\sqrt{75x^2 y^2 z^2}$

19) $\sqrt{4x^2 y^4 z^3}$

20) $\sqrt{12x^4 y^3 z^6}$

1) _____

2) _____

3) _____

4) _____

5) _____

6) _____

7) _____

8) _____

9) _____

10) _____

11) _____

12) _____

13) _____

14) _____

15) _____

16) _____

17) _____

18) _____

19) _____

20) _____

Score: _____

8-2 Solving Equations Involving Radicals

Solve each of the following. Check your answer by substituting back into the original equation.

1) $x^2 = 36$

2) $\dfrac{x}{9} = \dfrac{4}{x}$

3) $\dfrac{4x}{25} = \dfrac{4}{x}$

4) $2x^2 - 8 = 0$

5) $x^2 + 9 = 25$

6) $x^2 + 25 = 169$

7) $\sqrt{x} = 5$

8) $\sqrt{x + 3} = 5$

9) $\sqrt{5x + 1} = 4$

10) $\sqrt{8x + 1} = 5$

11) $3\sqrt{4x} = 12$

12) $\sqrt{7x} = 7$

13) $\sqrt{2x + 9} = 6$

14) $\sqrt{3x + 4} = 5$

15) $\sqrt{5x + 1} + 2 = 6$

16) $5\sqrt{2x} = 20$

17) $\sqrt{4x} + 2 = 6$

18) $\sqrt{x - 5} + 2 = 3$

19) $2\sqrt{3x - 15} = 6$

20) $\sqrt{2x - 3} = \sqrt{5x - 21}$

1)	
2)	
3)	
4)	
5)	
6)	
7)	
8)	
9)	
10)	
11)	
12)	
13)	
14)	
15)	
16)	
17)	
18)	
19)	
20)	

Score:

8-3 Adding and Subtracting Radical Expressions

Add or subtract each of the following. Express in simplest form.

1) $8\sqrt{5} + 3\sqrt{5}$

2) $7\sqrt{5} - 3\sqrt{5}$

3) $4\sqrt{3} + 6\sqrt{3} - 5\sqrt{3}$

4) $\sqrt{12} + 5\sqrt{3}$

5) $3\sqrt{6} - \sqrt{6} - 5\sqrt{6}$

6) $\sqrt{27} - 2\sqrt{3}$

7) $5\sqrt{8} + 15\sqrt{2}$

8) $\sqrt{18} + \sqrt{50}$

9) $\sqrt{45} - \sqrt{20}$

10) $\sqrt{4x} + \sqrt{81x^3}$

11) $6\sqrt{12} - \sqrt{75}$

12) $2\sqrt{12} + 5\sqrt{3}$

13) $2\sqrt{3} + \sqrt{12}$

14) $\sqrt{45} + 6\sqrt{20}$

15) $7\sqrt{8} - \sqrt{18}$

16) $6\sqrt{3} - 4\sqrt{27}$

17) $\sqrt{27} + \sqrt{75}$

18) $\sqrt{80} - \sqrt{5}$

19) $3\sqrt{50} - 5\sqrt{18}$

20) $5\sqrt{3} + 4\sqrt{12} - 2\sqrt{75}$

1) _____

2) _____

3) _____

4) _____

5) _____

6) _____

7) _____

8) _____

9) _____

10) _____

11) _____

12) _____

13) _____

14) _____

15) _____

16) _____

17) _____

18) _____

19) _____

20) _____

Score: _____

8-4 Multiplying Radical Expressions

Multiply and simplify each of the following.

1) $\sqrt{2} \cdot \sqrt{8}$

2) $\sqrt{3} \cdot \sqrt{8}$

3) $\sqrt{14} \cdot \sqrt{2}$

4) $2\sqrt{3} \cdot 5\sqrt{27}$

5) $6\sqrt{2} \cdot \sqrt{3}$

6) $5\sqrt{6} \cdot 2\sqrt{3}$

7) $\sqrt{10} \cdot \sqrt{20}$

8) $8\sqrt{2} \cdot \dfrac{\sqrt{14}}{8}$

9) $2\sqrt{18} \cdot 3\sqrt{8}$

10) $\sqrt{5} \left(\sqrt{5} + \sqrt{11} \right)$

11) $3\sqrt{2} \cdot 4\sqrt{18}$

12) $\sqrt{5} \left(\sqrt{5} - 2 \right)$

13) $6\sqrt{72}$

14) $3\sqrt{6} \cdot 5\sqrt{2}$

15) $\left(2\sqrt{3} \right)^2$

16) $\sqrt{3x} \cdot \sqrt{6x}$

17) $\left(5 + \sqrt{2} \right)\left(7 - \sqrt{2} \right)$

18) $\sqrt{2} \left(\sqrt{8} - 2\sqrt{2} + 6 \right)$

19) $\left(5 + \sqrt{3} \right)^2$

20) $\left(3 - \sqrt{6} \right)^2$

1) _____

2) _____

3) _____

4) _____

5) _____

6) _____

7) _____

8) _____

9) _____

10) _____

11) _____

12) _____

13) _____

14) _____

15) _____

16) _____

17) _____

18) _____

19) _____

20) _____

Score: _____

8-5 Dividing Radical Expressions

Divide and simplify each of the following.

1) $\dfrac{\sqrt{27}}{\sqrt{3}}$

2) $\dfrac{\sqrt{20}}{\sqrt{5}}$

3) $\dfrac{\sqrt{56}}{\sqrt{7}}$

4) $\dfrac{\sqrt{30a^3}}{\sqrt{6a^2}}$

5) $\dfrac{25\sqrt{24}}{5\sqrt{2}}$

6) $\dfrac{16\sqrt{24}}{4\sqrt{3}}$

7) $\dfrac{9\sqrt{x^3}}{3\sqrt{x}}$

8) $\dfrac{36\sqrt{36x^5}}{4\sqrt{4x^3}}$

9) $\dfrac{21\sqrt{40}}{\sqrt{5}}$

10) $\dfrac{\sqrt{6xx^3}}{2x}$

11) $\dfrac{8\sqrt{48}}{4\sqrt{2}}$

12) $\dfrac{\sqrt{x^3}}{\sqrt{x}}$

13) $\dfrac{\sqrt{21}+\sqrt{35}}{\sqrt{7}}$

14) $\dfrac{\sqrt{63y^3}}{\sqrt{7y}}$

15) $\dfrac{\sqrt{48x^3}}{\sqrt{3x}}$

16) $\dfrac{\sqrt{100x^3}}{\sqrt{25x}}$

17) $\dfrac{\sqrt{45x^4}}{\sqrt{9}}$

18) $\dfrac{12\sqrt{18x^4y^2}}{4\sqrt{2x^2}}$

19) $\dfrac{\sqrt{48x^2y^4}}{\sqrt{12xy^2}}$

20) $\dfrac{8\sqrt{12}}{2\sqrt{3}}$

1) _____

2) _____

3) _____

4) _____

5) _____

6) _____

7) _____

8) _____

9) _____

10) _____

11) _____

12) _____

13) _____

14) _____

15) _____

16) _____

17) _____

18) _____

19) _____

20) _____

Score: _____

8-6 Rationalizing the Denominator

Rationalize the denominator and simplify completely each of the following.

1) $\dfrac{\sqrt{5}}{\sqrt{3}}$

2) $\dfrac{12}{\sqrt{2}}$

3) $\dfrac{1}{\sqrt{2}}$

4) $\dfrac{5}{\sqrt{3}}$

5) $\dfrac{8}{\sqrt{2}}$

6) $\dfrac{25}{\sqrt{20}}$

7) $\dfrac{\sqrt{5}}{\sqrt{8}}$

8) $\dfrac{\sqrt{x^2}}{\sqrt{2}}$

9) $\dfrac{3\sqrt{18}}{\sqrt{2}}$

10) $\dfrac{\sqrt{10}}{\sqrt{3}}$

11) $\dfrac{\sqrt{2}}{\sqrt{6}}$

12) $\dfrac{\sqrt{x^2}}{\sqrt{8}}$

13) $\dfrac{\sqrt{13}}{\sqrt{2}}$

14) $\sqrt{\dfrac{3}{2}}$

15) $\dfrac{\sqrt{40}}{\sqrt{5}}$

16) $\dfrac{3}{\sqrt{12}}$

17) $\dfrac{15\sqrt{3}}{\sqrt{5}}$

18) $\dfrac{60}{3\sqrt{8}}$

19) $\dfrac{18}{3\sqrt{2}}$

20) $\dfrac{4}{\sqrt{12}}$

1) _____

2) _____

3) _____

4) _____

5) _____

6) _____

7) _____

8) _____

9) _____

10) _____

11) _____

12) _____

13) _____

14) _____

15) _____

16) _____

17) _____

18) _____

19) _____

20) _____

Score: _____

8-7 Simplifying Radical Expressions with Binomial Denominators

Rationalize the denominator and simplify each of the following. Remember to express your answers in their simplest form. You may need to cancel out common factors.

1) $\dfrac{3}{4 + \sqrt{5}}$

2) $\dfrac{\sqrt{10}}{3 + \sqrt{6}}$

3) $\dfrac{20}{\sqrt{6} + 2}$

4) $\dfrac{24}{\sqrt{15} - 3}$

5) $\dfrac{26}{4 - \sqrt{3}}$

6) $\dfrac{45}{\sqrt{2} + \sqrt{7}}$

7) $\dfrac{48}{\sqrt{11} - \sqrt{5}}$

8) $\dfrac{\sqrt{3} - 2\sqrt{5}}{\sqrt{5} - \sqrt{3}}$

9) $\dfrac{\sqrt{13} + 3}{5 - \sqrt{13}}$

10) $\dfrac{\sqrt{3} + 6}{\sqrt{3} + 5}$

11) $\dfrac{\sqrt{10} - \sqrt{6}}{\sqrt{10} + \sqrt{6}}$

12) $\dfrac{\sqrt{7} + \sqrt{5}}{\sqrt{7} - \sqrt{5}}$

13) $\dfrac{1}{1 + \sqrt{5}}$

14) $\dfrac{3}{\sqrt{5} - 2}$

15) $\dfrac{5}{2 + \sqrt{3}}$

16) $\dfrac{1 + \sqrt{7}}{2 - \sqrt{7}}$

17) $\dfrac{\sqrt{7}}{5 + \sqrt{3}}$

18) $\dfrac{\sqrt{8}}{4 - \sqrt{2}}$

19) $\dfrac{4\sqrt{3}}{5 - \sqrt{2}}$

20) $\dfrac{10}{2 + \sqrt{3}}$

1) _____

2) _____

3) _____

4) _____

5) _____

6) _____

7) _____

8) _____

9) _____

10) _____

11) _____

12) _____

13) _____

14) _____

15) _____

16) _____

17) _____

18) _____

19) _____

20) _____

Score: _____

8-8 The Pythagorean Theorem

Find the length of the missing side. Use a calculator that has a square root function. Round answers to nearest hundredth.

1)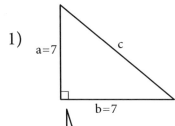

2)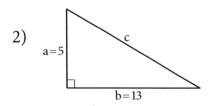

3)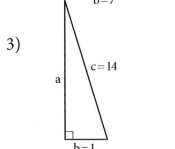

4)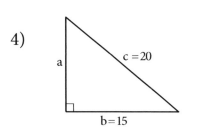

5)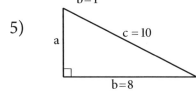

6)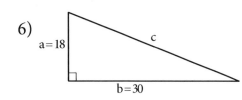

7) a = 16, b = 63, c = ?   8)   a = 16, b = ?, c = 34

9) a = 6, b = 3, c = ?   10) a = ?, b = 10, c = 25

11) a = ?, b = 20, c = 29   12) a = 7, b = 24, c = ?

13) a = 2, b = 6, c = ?   14) a = 9, b = ?, c = 15

15) a = ?, b = 5, c = 11   16) a = 5, b = 7, c = ?

17) a = 8, b = 15, c = ?   18) a = 6, b = ?, c = 10

19) a = ?, b = 15, c = 17   20) a = 25, b = 30, c = ?

1) _____
2) _____
3) _____
4) _____
5) _____
6) _____
7) _____
8) _____
9) _____
10) _____
11) _____
12) _____
13) _____
14) _____
15) _____
16) _____
17) _____
18) _____
19) _____
20) _____

Score: _____

8-9 The Distance Formula

Draw a sketch and use the Distance Formula to find the distance between the given points. Round each answer to the nearest tenth.

$$D = \sqrt{(x_2 - x_1)^2 + (y_2 - y_1)^2}$$

1) (2, 3) (-4, 6)

2) (-5, 4) (3, -2)

3) (-6, 0) (4, 0)

4) (3, 4) (9, 7)

5) (-5, -3) (-9, -6)

6) (-4, 6) (5, 2)

7) (4, -4) (9, -8)

8) (-2, 1) (-8, -5)

9) (2, 2) (5, 6)

10) (8, -5) (3, 7)

11) (4, 2) (-2, 10)

12) (-5, 1) (7, 6)

13) (4, -2) (1, 2)

14) (-2, 4) (4, -2)

15) (3, 1) (-2, -1)

16) (-2, 4) (7, -8)

17) (2, -1) (5, 3)

18) (1, -6) (6, 6)

19) (8, 10) (2, 2)

20) (6, -18) (-4, 6)

1) _____

2) _____

3) _____

4) _____

5) _____

6) _____

7) _____

8) _____

9) _____

10) _____

11) _____

12) _____

13) _____

14) _____

15) _____

16) _____

17) _____

18) _____

19) _____

20) _____

Score: _____

Chapter 8: **RADICAL EXPRESSIONS AND GEOMETRY**

8-10 The Midpoint Formula

Find the coordinates of the midpoint between the two given points.

$$M = (\frac{x_1 + x_2}{2}), (\frac{y_1 + y_2}{2})$$

1) (7, 1) (-3, 1)

2) (5, -2) (9, -8)

3) (-4, 4) (4, -4)

4) (6, 8) (2, 4)

5) (-2, -8) (-6, -2)

6) (-3, -4) (6, -8)

7) (-3, 3) (1, 0)

8) (6, -8) (2, 4)

9) (-2, -1) (-8, 6)

10) (7, 3) (-1, -3)

11) (3, -5) (-5, 11)

12) (2, 1) (-9, 3)

13) (10, 9) (-5, 6)

14) (-1, -3) (-5, -7)

15) (-4, 8) (8, -2)

16) (2, 2) (-8, -7)

17) (5, 3) (-1, 0)

18) (6, -2) (9, -6)

19) (3, -2) (-5, -2)

20) (3, -2) (-5, -2)

1) _____

2) _____

3) _____

4) _____

5) _____

6) _____

7) _____

8) _____

9) _____

10) _____

11) _____

12) _____

13) _____

14) _____

15) _____

16) _____

17) _____

18) _____

19) _____

20) _____

Score: _____

9-1 Factoring Quadratic Equations

Solve each of the following.

1) $x^2 + 8x + 15 = 0$ 2) $x^2 - 4x = 5$

3) $x^2 + 2x = 15$ 4) $x^2 + 3x = 0$

5) $x^2 = 8x$ 6) $x^2 - 4x = 12$

7) $x^2 - 81 = 0$ 8) $x^2 + 8x = 0$

9) $\dfrac{x}{3} = \dfrac{8}{x+2}$ 10) $x = \dfrac{40}{x-3}$

11) $x^2 = x + 72$ 12) $\dfrac{x}{3} + \dfrac{9}{x} = 4$

13) $x^2 - 7x = -10$ 14) $2x^2 = 3x$

15) $x^2 = 15 - 2x$ 16) $x^2 - 8x + 28 = 3x$

17) $x^2 = -3x$ 18) $x(x - 2) = 35$

19) $x^2 - 8x + 12 = 0$ 20) $x^2 + 10x = -9$

1) _____
2) _____
3) _____
4) _____
5) _____
6) _____
7) _____
8) _____
9) _____
10) _____
11) _____
12) _____
13) _____
14) _____
15) _____
16) _____
17) _____
18) _____
19) _____
20) _____

Score: _____

9-2 Solving Quadratic Equations Using Square Roots

Solve each of the following.

1) $x^2 - 144 = 0$ 2) $x^2 = 36$

3) $3x^2 = 147$ 4) $5x^2 - 5 = 0$

5) $x^2 - 6 = 58$ 6) $\dfrac{4x}{25} = \dfrac{4}{x}$

7) $2x^2 - 11 = 39$ 8) $(x - 3)^2 = 49$

9) $\dfrac{2x}{9} = \dfrac{6}{x}$ 10) $\dfrac{x}{8} = \dfrac{4}{x}$

11) $x^2 - 25 = 25$ 12) $4x^2 - 14 = 2x^2$

13) $(x - 1)^2 = 4$ 14) $(x + 5)^2 = 23$

15) $7x^2 = 3x^2 + 36$ 16) $3x^2 = 6$

17) $x^2 + 25 = 100$ 18) $4x^2 = 48$

19) $(x - 2)^2 = 3$ 20) $2(4x - 2)^2 = 72$

1) _____
2) _____
3) _____
4) _____
5) _____
6) _____
7) _____
8) _____
9) _____
10) _____
11) _____
12) _____
13) _____
14) _____
15) _____
16) _____
17) _____
18) _____
19) _____
20) _____

Score: _____

9-3 Completing the Square

Solve each of the following by completing the square.

1) $x^2 + 2x - 8 = 0$

2) $x^2 - 4x - 21 = 0$

3) $x^2 - 4x + 3 = 0$

4) $x^2 - 6x - 40 = 0$

5) $x^2 - 4x - 7 = 0$

6) $x^2 + 8x + 15 = 0$

7) $x^2 + 14x - 15 = 0$

8) $x^2 - 4x - 11 = 0$

9) $x^2 - 6x + 8 = 0$

10) $x^2 - 4x + 2 = 0$

11) $x^2 - 4x - 1 = 0$

12) $x^2 - 3x - 28 = 0$

13) $x^2 - 4x - 12 = 0$

14) $x^2 + 3x - 10 = 0$

15) $x^2 - 4x = 21$

16) $x^2 + 4x + 3 = 0$

17) $x^2 - 8x + 7 = 0$

18) $x^2 - 6x + 4 = 0$

19) $x^2 + 2x - 1 = 0$

20) $x^2 - 6x - 20 = 0$

1) _____

2) _____

3) _____

4) _____

5) _____

6) _____

7) _____

8) _____

9) _____

10) _____

11) _____

12) _____

13) _____

14) _____

15) _____

16) _____

17) _____

18) _____

19) _____

20) _____

Score: _____

Chapter 9: **QUADRATIC EQUATIONS**

9-4 The Quadratic Formula

Use the quadratic formula to solve each of the following.

$$x = \frac{-b \pm \sqrt{b^2 - 4ac}}{2a}$$

1) $x^2 - 5x + 6 = 0$

2) $3x^2 - 7x + 2 = 0$

3) $x^2 - 5x + 1 = 0$

4) $x^2 + x - 20 = 0$

5) $x^2 - x - 5 = 0$

6) $2x^2 - x - 4 = 0$

7) $3x^2 - 2x - 8 = 0$

8) $x^2 - 10x + 22 = 0$

9) $x^2 - 4x - 5 = 0$

10) $x^2 - 7x + 10 = 0$

11) $3x^2 + 4x + 1 = 0$

12) $5x^2 - 8x + 3 = 0$

13) $x^2 - 2x - 3 = 0$

14) $x^2 - 2x - 5 = 0$

15) $x^2 + 2x - 1 = 0$

16) $x^2 - 2x - 2 = 0$

17) $x^2 - 4x - 21 = 0$

18) $2x^2 - 5x - 1 = 0$

19) $2x^2 - 6x - 10 = 0$

20) $2x^2 - 3x - 4 = 0$

1) _____

2) _____

3) _____

4) _____

5) _____

6) _____

7) _____

8) _____

9) _____

10) _____

11) _____

12) _____

13) _____

14) _____

15) _____

16) _____

17) _____

18) _____

19) _____

20) _____

Score: _____

9-5 Graphing Quadratic Equations

Use the 4-step method to graph each of the following.
Be sure to refer to Lesson 9-5 in the text.

1) $y = x^2 - 5$

2) $y = -x^2 + 4x + 5$

3) $y = x^2 + 4x - 9$

4) $y = -x^2 + 5x + 6$

5) $y = x^2 - 5$

6) $y = x^2 + 4x - 9$

7) $y = -x^2 + 7$

8) $y = 3x^2 + 6x + 3$

9) $y = x^2 - 6x + 5$

10) $y = x^2 - 4x - 4$

11) $y = x^2 - 2x - 8$

12) $y = -3x^2 - 6x + 4$

13) $y = 4x^2$

14) $y = x^2 + 2$

15) $y = -x^2 + 2x + 3$

16) $y = x^2 - 4x + 3$

17) $y = -2x^2$

18) $y = 3x^2 - 6x + 4$

19) $y = -x^2 + 5$

20) $y = -3x^2 + 6x + 1$

1) _____

2) _____

3) _____

4) _____

5) _____

6) _____

7) _____

8) _____

9) _____

10) _____

11) _____

12) _____

13) _____

14) _____

15) _____

16) _____

17) _____

18) _____

19) _____

20) _____

Score: _____

9-6 The Discriminant

Determine how many times the graph of each quadratic function passes through the x-axis by using the discriminant **b² – 4ac**.

- If $b^2 - 4ac < 0$, the graph crosses the x-axis 0 times
- If $b^2 - 4ac > 0$, the graph crosses the x-axis 2 times
- If $b^2 - 4ac = 0$, the graph crosses the x-axis 1 time

1) $y = x^2 + 5x - 6$

2) $y = x^2 + 8x + 16$

3) $y = 2x^2 + 10x + 11$

4) $y = 3x^2 + 4x + 2$

5) $y = 3x^2 + 2x - 6$

6) $y = x^2 - 5x + 9$

7) $y = 25x^2 + 30x + 9$

8) $y = x^2 - 5x + 4$

9) $y = x^2 - 16x + 64$

10) $y = 3x^2 - 4x + 3$

11) $y = -5x^2 + x + 1$

12) $y = x^2 - x + 2$

13) $y = 3x^2 + 2x - 7$

14) $y = x^2 - 5x + 7$

15) $y = x^2 + 12x + 36$

16) $y = 4x^2 - 3x + 3$

17) $y = 3x^2 - 3x + 4$

18) $y = 3x^2 + 5x + 7$

19) $y = 3x^2 + 10x + 2$

20) $y = x^2 + 3x - 5$

1)
2)
3)
4)
5)
6)
7)
8)
9)
10)
11)
12)
13)
14)
15)
16)
17)
18)
19)
20)

Score:

10-1 Introduction to Algebra Word Problems

Translate each of the following into an equation

1) Twelve more than three times a number is 50.

2) Forty less than twice a number is equal to 90.

3) The sum of a number and 15 is twelve.

4) Fifteen is ten less than four times a number.

5) Five times a number less six equals eight more than twice the number.

6) Twice the sum of a number and six is eighteen.

7) The sum of five times a number and 12 is 95.

8) The difference between four times a number and 7 equals -20.

9) Twice a number increased by 20 equals 48.

10) Four more than three times a number is seven less that five times the number.

11) The product of a number and three is sixty-three.

12) The quotient of a number and 6 equals five.

13) Thirty-six equals a number less six

14) Fifteen added to five times a number equals three less than seven times the number.

15) Twice the quantity of a number added to five is equal to 30.

16) Six more than twice a number equals that number less thirty.

17) The difference of five times a number and 7 is thirty-four.

18) The sum of twelve and one-third a number is 42.

19) Five times a number decreased by seven equals nine.

20) Six times the difference of a number and eight is equal to fifteen.

1) _____

2) _____

3) _____

4) _____

5) _____

6) _____

7) _____

8) _____

9) _____

10) _____

11) _____

12) _____

13) _____

14) _____

15) _____

16) _____

17) _____

18) _____

19) _____

20) _____

Score: _____

10-2 Everyday Algebra Word Problems

Solve each algebra problem using a variable and an equation.

1) The difference between seven times a number and 23 is 82. Find the number.

2) Ten times a number added to 9 is 59. Find the number

3) Twelve times a number less 6 is ninety. Find the number.

4) Two-thirds a number is 40. Find the number.

5) Nine more than one-half a number is twenty-nine. Find the number.

6) John is five times as old as Mike. If the sum of their ages is 18, what is the age of each boy?

7) One number is 12 greater than the other. If their sum is 36, find the numbers

8) One number is five times greater than another. If their difference is 96, find the numbers.

9) The number of boys enrolled in a school is 235 more than the number of girls. If the total number of students enrolled is 885. How many boys are enrolled, and how many girls are enrolled?

10) Find two consecutive integers whose sum is 251.

11) Find three consecutive integers whose sum is 270.

12) Five times a number less seven is eight more than twice the number. Find the number.

13) Find three consecutive even integers whose sum is 378.

14) Twice the sum of a number and 16 is 50. Find the number.

15) One number is 6 time another. If their sum is 70, find the numbers.

1) _____

2) _____

3) _____

4) _____

5) _____

6) _____

7) _____

8) _____

9) _____

10) _____

11) _____

12) _____

13) _____

14) _____

15) _____

Score: _____

10-3 Time, Rate, and Distance Problems (continued)

Solve each of the following. Identify the type, draw a sketch, and use a chart. Be sure to refer to lesson 10-3 in the *No-Nonsense Algebra* text.

1) Joe and Mary each leave home driving in opposite directions. If Joe travels at the speed of 56 mph, and Mary travels at the speech of 70 mph. In how many hours will they be 504 miles apart?

2) Two trains are 800 miles apart. They travel toward each other, one at 67 mph and the other at 53 mph. How much time will it take before they meet?

3) Ray left his house traveling at a speed 45 mph. Two hours later, Steve left the same house trying to catch up with Ray. If he was traveling at the speed of 60 mph, how long would it take for him to catch up with Ray?

4) Paula spent 6 hours on a hike. She left the parking lot and walked to the top of a mountain at a rate of 2 mph, and back down the mountains to the parking lot at a rate of 4 mph. How far was it to the top of the mountain?

5) Two cars started from the same point, driving in opposite directions. One traveled at a rate of 40 mph, and the other at a rate of 30 mph. In how many hours would they be 350 miles apart?

6) Two trains are 700 miles apart. They start traveling toward each other. One travels at a rate of 65 mph, and the other at a rate of 75 mph. How much time would it take until they meet?

7) Mr. Smith left his house traveling at a rate of 40 mph. One hour later, his son, Ron, left the house attempting to catch up with his father. The son was traveling at a rate of 50 mph. How long will it take for Ron to catch up with Mr. Smith?

8) Two planes leave the same airport at the same time, traveling in opposite directions. One traveled at the rate of 450 mph, and the other at the rate of 600 mph. In how many hours will they be 6,300 miles apart?

9) Two cars are 360 miles apart. They travel toward each other, one at 70 mph, and the other at 50 mph. How much time will it take before they meet?

10) Susan and Jane are taking a motorcycle trip. Susan gets a 1 hour head start. She is traveling at a rate of 40 mph. Her friend, Jane, leaves to catch up traveling at a rate of 55 mph. How much time will it take Jane to catch up with Susan?

1)
2)
3)
4)
5)
6)
7)
8)
9)
10)
Score:

10-4 Mixture Problems

Use a chart to solve each of the following problems. Refer to lesson 10-4 in your *No-Nonsense Algebra* text.

1) A store sells mixed nuts worth $2.40 per pound and mixed fruit worth $1.50 per pound. How many pounds of each is needed to produce a mixture of 30 pounds to sell for $2.25 per pound?

2) If cashews sell for $3.60 per pound and walnuts sell for $2.25 per pound, how many pounds of each must be used to produce 45 pounds of a mixture to sell at $3.00 per pound?

3) A store has some chocolate candy that sells for $1.80 per pound and some vanilla candy that sells for $2.80 per pound. The store wants 120 pounds of a mixture that sells for $2.20 per pound. How many pounds of each kind should be mixed?

4) A baker makes peanut butter cookies worth $2.85 per pound and pecan cookies worth $5.10 per pound. How many pounds of each kind must he used to produce a 45-pound mixture to sell for $3.75 per pound?

5) Peanuts sell for $3.50 per pound and cashews sell for $4.75 per pound. How many pounds of each are needed to make 20 pounds of a mixture that will sell for $4.00 per pound?

6) A farmer has some cream which is 24% butterfat and some cream which is 18% butterfat. How many quarts of each must be used to produce 90 pounds of cream which is 22% butterfat?

7) How many liters of a solution which is 75% pure acid must be mixed with 16 liters of a solution which is 30% pure acid to produce a solution which is 55% pure acid?

8) A scientist has 200 liters of a solution that is 40% acid. How many liters of water should be added to make a solution that is 25% acid?

9) A salt solution contains 6% salt. How much water must be added to 10 liters of the solution to get a mixture that only 4% salt?

10) A scientist needs 70 liters of a 50% acid solution. He has an 80% acid solution and a 30% acid solution. How many liters of each must he mix to produce the 70 liters of 50% acid solution?

1)
2)
3)
4)
5)
6)
7)
8)
9)
10)
Score:

10-5 Work Problems

Solve each of the following. Use a chart.

1) John can paint a room in 6 hours. Mary can paint the same room in 9 hours. How long will it take to paint the room if they work together?

2) Ralph can clean the house in 2 hours. Johan can clean it in 3 hours. How long will it take to clean the house if they work together?

3) June can wax the hardwood floors in 4 hours, and Steve can wax the floors in 3 hours. How long will it take if they work together?

4) Allen can paint a shed in 5 hours. If he and Ed work together, they can paint it in 2 hours. How long will it take for Ed to paint the shed alone?

5) Jane can type a 50-page paper in 8 hours. If Jane and Mary can type the paper in 6 hours working together, how long would it take Mary to type the paper working alone?

6) Company A can install a new roof in 10 hours. Company B can install the roof in 15 hours. How long will it take to install the roof if they work together?

7) A pump can fill a pool in 10 hours. Another pump can empty the pool in 20 hrs. With both pumps are working, how long will it take to fill the pool?

8) One inlet pipe can fill a tank in 3 hours. An outlet pipe can empty the tank in 6 hours. If the tank is empty and both pipes are open, how long will it take to fill the tank?

9) John can paint a fence in 12 hours. His son can paint the fence in 24 hours. How long will it take to paint the fence if they work together?

10) Company X can install carpet in 26 hours. Company Y can install the carpet in 39 hours. How long would it take if they worked together?

1) _____

2) _____

3) _____

4) _____

5) _____

6) _____

7) _____

8) _____

9) _____

10) _____

Score: _____

10-6 Age Problems

Solve each of the following.

1) John is 3 times as old as Tom. In 18 years from now, John will be twice as old as Tom will be then. Find the present age of both.

2) A father is 3 times as old as his son. Fifteen years ago, the father was 9 times as old as his son was then. Find their present ages.

3) Jim is 24 years older than Roger. Four years ago, he was 7 times as old as Roger was then. Find their present ages.

4) Jose is 20 years older than Maria. Sixteen years ago, Jose was 3 times as old as Maria was then. Find their present ages.

5) Brian is 15 years older than Lee. Five years from now, Brian will be 1½ times as old as Lee will be then. How old is each of them now?

6) Joe is now 3 times as old as Cal. In ten years, Joe will be twice as old as Cal will be then. Find their present ages.

7) William is one-half the age of his father. Twelve years ago, William was one-third as old as his father was then. Find their present ages.

8) Julie is 25 years older than Kim. Ten years from now, Julie will be twice as old as Kim. What are their present ages?

9) A mother is 4 times as old as her daughter. Three years from now, the mother will be 3 times as old as the daughter is then. What are their present ages?

10) Elena is 6 years older than Katya. Six year ago, Elena was twice as old as Katya. What are their present ages?

1)
2)
3)
4)
5)
6)
7)
8)
9)
10)
Score:

10-7 Coin Problems

Solve each of the following.

1) John has 3 times as many dimes as nickels. The value of the coins is $1.40. How many of each type of coin does he have?

2) A girl has twice as many dimes as pennies, and 3 times as many nickels as pennies. The total value is $1.80. How many coins of each type does she have?

3) Ashley has three times as many dimes as nickels and twice as many quarters as dimes. The total value is $5.55. How many coins of each does she have?

4) Leigh has $2.05 in quarters and dimes. She has 4 more quarters than dimes. How many of each type of coin does she have?

5) Amir has a collection of 103 dimes and quarters. The total value is $15.25. How many of each type are there?

6) A jar of quarters and nickels has a total value of $1.25. There are 13 coins in all. How many of each type are there?

7) Jim has a jar of nickels and dimes. There are 5 times as many dimes as nickels. If the total value is $4.40, how many nickels are there? How many dimes are there?

8) Stewart has some nickels and dimes. The total value is $1.65. There are 12 more nickels than dimes. How many of each kind of coin does he have?

9) A boy has a collection of nickels, dimes, and quarters whose total value is $3.20. He has 3 times as many quarters as nickels, and 5 more dimes than nickels. How many of each type of coin are there?

10) Joe has $1.35 in nickels and dimes. Altogether there are 15 coins. How many of each type of coin are there?

1)	
2)	
3)	
4)	
5)	
6)	
7)	
8)	
9)	
10)	
Score:	

10-8 Investment Problems

Solve each of the following. Be sure to make a chart and refer to Lesson 10-8 in your *No-Nonsense Algebra* text. You may use a calculator.

1) A man invested some money at 6% yearly interest. He also invested $800 at 7%. If the money earned $128 interest in a year, how much was invested at 6%?

2) Mr. Smith invested some money 6% and a second sum that was $500 more than the first, at 7%. The total interest earned was $555. How much did he invest at each rate?

3) Cheryl invested some money at 6%. She also invested twice as much at 5%. The total interest earned in a year was $680. How much did she invest at each rate?

4) Mr. and Mrs. Johnson invested a sum of money at 6%. They also invested a sum that was $250 more than the first sum, at 8%. If the total annual interest was $300, how much did they invest at each rate?

5) A family invested a sum of money at 6%. They also invested a second sum that was $2000 less than the first sum at 10%. Find the amount invested at each rate if the annual interest earned was $620.

6) Bob invested $1600 at 7% interest. He also invested a second sum at 8.5%. The interest earned in a year was $197. How much money was invested at 8.5%?

7) Ms. Allen invested $1200. Part was invested at 6% and the rest at 8%. If the interest earned in one year was $78, how much was invested at each rate?

8) A total of $3500 is invested in two parts. One part is invested at a rate of 6% and the second part is invested at the rate of 8%. If the total interest earned in a year was $250, find the amount invested at each rate.

9) Dave invested $2000. Part was invested at the rate of 6% and the rest at 8%. If the interest earned in a year was $130, how much did he invest at each rate?

10) Mr. Jones inherited $2,000,000 and decided to invest it. He invested $750,000 at 8%, and the rest, $1,250,000 at 12%. How much interest will he earn in one year?

1) _____

2) _____

3) _____

4) _____

5) _____

6) _____

7) _____

8) _____

9) _____

10) _____

Score: _____

Final Review

Simplify 1–20.

1) $-7 + 9 - 6 - -7$

2) $(-6)(-5)(3)$

3) $\dfrac{-108}{6}$

4) $\dfrac{-80 \div 20}{2 \div -2}$

5) $-\dfrac{2}{3} + \dfrac{1}{4}$

6) $-\dfrac{3}{8} - \dfrac{1}{4}$

7) $-2\dfrac{1}{2} \cdot \dfrac{3}{4}$

8) $-4\dfrac{3}{4} \div -\dfrac{1}{4}$

9) $-5.12 + 7.6$

10) $-3.6 - -12$

11) $6(-7.82)$

12) $\dfrac{-6.3}{.09}$

13) 8^4

14) $\dfrac{\sqrt{225}}{\sqrt{25}}$

15) $4^3 \cdot 4^2$

16) $6 + \{(5 + 6)3\}$

17) $85 \div 5 - 4 \times 2 + 8$

18) $5(4^2 + 2) - 5^2$

19) $\dfrac{5^2 + 15}{2^3}$

20) $5^2 + \{(3^2 + 2^2)3\}$

1)	
2)	
3)	
4)	
5)	
6)	
7)	
8)	
9)	
10)	
11)	
12)	
13)	
14)	
15)	
16)	
17)	
18)	
19)	
20)	
Score:	

21) Give a prime factorization for 180.

22) Find the greatest common factor of 300 and 225.

23) Find the least common multiple of 36 and 40.

24) Write 425,000,000,000 in scientific notation.

25) Write .000096 in scientific notation.

26) Solve the proportion $\frac{8}{6x} = \frac{4}{5}$

27) Find 90% of 650.

28) 120 is 75% of what?

29) 190 is what % of 760?

21) _____

22) _____

23) _____

24) _____

25) _____

26) _____

27) _____

28) _____

29) _____

Score: _____

Final Review

For 30 through 44, solve each equation.

30) $x - 13 = 45$

31) $x - -12 = -73$

32) $\frac{3}{4}y = 27$

33) $\frac{x-7}{4} = 11$

34) $\frac{1}{2}n = 7\frac{1}{2}$

35) $5x - 9 = -3x + 7$

36) $10x - 2 = 8x - 1$

37) $4(3n + 5) = 8$

38) $2(m - 3) + 5 = 3(m-1)$

39) $|x - 4| = 3$

40) $|4x| = 20$

41) $\frac{3x-2}{5} = \frac{7}{10}$

42) $\frac{4n-8}{-2} = 12$

43) $\frac{x}{5} + 6 = -2$

44) $8x + 18 = -3(x + 5)$

30) _____

31) _____

32) _____

33) _____

34) _____

35) _____

36) _____

37) _____

38) _____

39) _____

40) _____

41) _____

42) _____

43) _____

44) _____

Score: _____

45) Sketch the graph of $y = 3x + 2$.

46) Sketch the graph of $y + 3x = 2$.

47) Find the x and y intercepts for $4x + 3y = 12$.

48) Find the slope of the line that passes through the points $(-2, -8)$ and $(1, 4)$.

49) Find the slope of the line that passes through the points $(18, -4)$ and $(6, -10)$.

50) Solve for y to find the slope of the equation $6x + 2y = 12$.

51) Find the equation of the line in slope-intercept form that has slope $= 4$, y-intercept $= -5$.

52) Find the equation of the line in slope-intercept form that has slope $= -2$ and passes through $(0, 6)$.

53) Find the equation of the line in slope-intercept form that passes through the points $(-2, -5)$ and $(-1, -2)$.

45) _____

46) _____

47) _____

48) _____

49) _____

50) _____

51) _____

52) _____

53) _____

Score: _____

54) Find the point-slope form of a line that passes through (4, 6) and has slope = 5.

55) Write the equation $2y = -3x + 6$ in standard form.

56) Sketch the graph of the inequality $x \geq 5$.

57) Sketch the graph of the inequality $-7 < x \leq 5$.

58) Sketch the graph of the inequality $x \leq -5$ or $x > 3$.

59) Sketch the graph of the inequality $|x| < 5$.

60) Sketch the graph of the inequality $|x| \geq 6$.

Solve each inequality.

61) $x + 5 \leq 7$

62) $-5x \geq 275$

63) $\frac{x}{8} - 13 > -6$

64) $|x + 4| \leq 3$

65) $|x + 5| > 7$

66) $2|x| > 16$

67) $4|x| < 20$

54) _____

55) _____

56) _____

57) _____

58) _____

59) _____

60) _____

61) _____

62) _____

63) _____

64) _____

65) _____

66) _____

67) _____

Score: _____

68) Sketch the graph of $y > x$.

69) Sketch the graph of $y \leq 2x + 2$.

70) Solve by using the substitution method.

$$y = 5x$$
$$x + 2y = 22$$

71) Solve by using the substitution method.

$$x = 2y + 3$$
$$3x + 4y = -1$$

72) Solve by using the elimination method.

$$3x + 4y = 6$$
$$5x + 2y = -4$$

73) Solve by using the elimination method.

$$4x - 3y = 12$$
$$x + 2y = 14$$

74) Sketch the graph that shows the solution set of the following system of inequalities.

$$x < -4$$
$$y > 3$$

75) Add $(-7x^2 + 4xy - 6y^2) + (-5x^2 - 12xy + 3y^2)$

76) Add $(3x^2 - 4x + 8) + (-7x^2 + 2x - 5)$

77) Subtract $(5x - 3) - (-2x + 1)$

68) _____
69) _____
70) _____
71) _____
72) _____
73) _____
74) _____
75) _____
76) _____
77) _____
Score: _____

Simplify each of the following.

78) $(6x^2 - 7xy - 4y^2) - (2x^2 + 5xy + 6y^2)$

79) $(-7x^3y^4)(4xy^3)$

80) $\dfrac{36x^3y^2}{5} \cdot \dfrac{35x^2y^4}{4}$

81) $\dfrac{18x^3y^4z^7}{-2x^2yz}$

82) $\dfrac{9x^5y^5z^7}{3x^2y^3z^5}$

83) $-2x^2(3x^2 - 7x + 10)$

84) $5y(-2y^2 - 7y)$

85) $(n + 6)(n + 7)$

86) $(4x + 3)(9x + 6)$

87) $\dfrac{4x^3 + 2x^2 - 5}{2x}$

88) $\dfrac{33x^4 + 11x^3 - 44x^2}{11x}$

78)	
79)	
80)	
81)	
82)	
83)	
84)	
85)	
86)	
87)	
88)	
Score:	

Final Review

Factor each of the following.

89) $15x^2y - 30xy$

90) $x^3 + 11x^2 + 24x$

91) $x^2 - 4x - 45$

92) $100x^2 - 36y^2$

93) $7x^2 - 7y^2$

94) $50x^2 + 100x - 400$

Simplify each of the following.

95) $\dfrac{7x + 7y}{8x + 8y}$

96) $\dfrac{x^2 - 25}{x^2 + 10x + 25}$

97) $\dfrac{3x + 6}{x^2 + 7x + 10}$

98) $\dfrac{x^2 - 9x + 8}{x^2 - 16x + 64}$

99) $\dfrac{x - 2}{x^2 - 4}$

100) $\dfrac{x^2 - 4}{x^2 + 4x + 4}$

101) $\dfrac{x - 3}{3 - x}$

102) $\dfrac{x - 5}{x} \bullet \dfrac{x^2}{x^2 - 2x - 15}$

103) $\dfrac{6x - 2}{7} = \dfrac{5x + 7}{8}$

89) _____

90) _____

91) _____

92) _____

93) _____

94) _____

95) _____

96) _____

97) _____

98) _____

99) _____

100) _____

101) _____

102) _____

103) _____

Score: _____

104) Divide $\dfrac{4x^3}{y^4} \div \dfrac{8x^2}{y^2}$

105) Divide $\dfrac{3x + 12}{4x - 18} \div \dfrac{2x + 8}{x + 4}$

106) Add $\dfrac{5x + 1}{2x + 1} + \dfrac{3x - 2}{2x + 1}$

107) Subtract $\dfrac{3x + 4}{x - 2} - \dfrac{x - 1}{x - 2}$

108) Add $\dfrac{3}{x^2} + \dfrac{5}{x}$

109) Add $\dfrac{x + 1}{x} + \dfrac{x - 3}{3x}$

110) Add $\dfrac{9}{4x} + \dfrac{3}{2x}$

111) Add $\dfrac{x + 9}{2} + \dfrac{x - 3}{3}$

112) Subtract $\dfrac{6x + y}{30} - \dfrac{2x - y}{10}$

113) Subtract $\dfrac{x + 7}{3} - \dfrac{2x - 3}{5}$

114) Solve $\dfrac{x + 2}{4} = \dfrac{5}{2}$

104) _____

105) _____

106) _____

107) _____

108) _____

109) _____

110) _____

111) _____

112) _____

113) _____

114) _____

Score: _____

115) Solve $\quad \frac{4}{x} + \frac{1}{x} = \frac{3}{4}$

116) Solve $\quad \frac{2}{x} - \frac{3}{4x} = \frac{1}{3}$

Simpify each.

117) $\sqrt{75}$

118) $\sqrt{18x^3}$

119) $\sqrt{80x^2}$

120) $\sqrt{144x^4 y^2}$

For 121 - 123, rationalize the denominator.

121) $\frac{7}{\sqrt{5}}$

122) $\frac{\sqrt{18}}{\sqrt{2}}$

123) $\frac{\sqrt{75}}{\sqrt{15}}$

Simpify each of the following.

124) $8\sqrt{5} + 5\sqrt{5}$

125) $\sqrt{2} + \sqrt{50}$

126) $3\sqrt{40} - \sqrt{90}$

127) $\sqrt{32} \cdot \sqrt{2}$

128) $\sqrt{5x^2 y} \cdot \sqrt{10y}$

115)
116)
117)
118)
119)
120)
121)
122)
123)
124)
125)
126)
127)
128)
Score:

129) Solve $\dfrac{x}{16} = \dfrac{4}{x}$

130) Solve $x^2 + 12 = 181$

131) Solve $x^2 - 12 = 69$

132) Rationalize the denominator. $\dfrac{2}{6 - \sqrt{3}}$

133) Rationalize the denominator. $\dfrac{3}{5 - \sqrt{2}}$

134) Solve for c.
You may use a calculator.

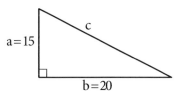

135) Solve for b.
You may use a calculator.

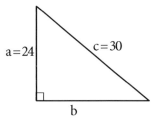

136) Use the Distance Formula to find the distance between the two points $(9, -2)$ and $(1, 13)$. You may use a calculator.

137) Use the Midpoint Formula to find the midpoint between the two points $(8, 3)$ and $(16, 7)$. You may use a calculator.

129) _____

130) _____

131) _____

132) _____

133) _____

134) _____

135) _____

136) _____

137) _____

Score: _____

138) Solve $5x^2 = 405$

139) Solve $3x^2 - 300 = 0$

140) Solve $x^2 - 3x = 28$

141) Solve $x^2 - 5x - 6 = 0$

142) Solve $x(x - 3) = 4$

143) Solve $(x + 4)^2 = 36$

144) Solve $(x - 6)^2 = 49$

145) Solve by completing the square. $x^2 - 5x - 14 = 0$

146) Solve by completing the square. $2x^2 - 4x = 1$

147) Solve by using the Quadratic Formula. $x^2 - 10x + 22 = 0$

148) Solve by using the Quadratic Formula. $x^2 - 4x + 1 = 0$

138) _____

139) _____

140) _____

141) _____

142) _____

143) _____

144) _____

145) _____

146) _____

147) _____

148) _____

Score: _____

149) Use the 4-step process to sketch a graph of the following.
$$y = x^2 + x + 3$$

150) Use the 4-step process to sketch a graph of the following.
$$y = -x^2 + 2x - 3$$

151) Six times a number less 10 is 14 greater than twice the number. Find the number.

152) Rhonda can ride her bike 2 km/hr faster than Ron. They both ride their bikes from the same house in opposite directions for two hours. They are then 172 km apart. How far does each of them travel?

153) Two kinds of nuts are mixed to sell at $4.00 per pound. Pistachios sell for $4.40 per pound, and pecans sell for $3.20 per pound. How much of each should be mixed to get a mixture that weighs 120 pounds?

154) Roy can wash a car in 2 hours. If his brother, Mike, can wash a car in 3 hours, how long will it take to wash a car if they work together?

155) A man invested $20,000. Part was invested at 5%, and the rest at 7%. The annual interest earned was $1,280. How much was invested at each rate?

156) A mother is 5 times as old as her daughter. Five years ago the mother was 15 times as old as her daughter. How old is each now?

157) A girl has a collection of nickels, dimes, and quarters. The total value is $4.15. There are 4 times as many dimes as nickels, and there are 3 less quarters than nickels. How many coins of each kind are there?

149) _____

150) _____

151) _____

152) _____

153) _____

154) _____

155) _____

156) _____

157) _____

Score: _____

Glossary

Abscissa The first number in an ordered pair that is assigned to a point on a coordinate plane. Also called the x-coordinate.

Absolute value The distance between 0 and a number on the number line. The absolute value of n is written $|n|$.

Algebra The branch of mathematics that uses letters and numbers to show the relationships between quantities.

Algebraic expression A mathematical expression which contains at least one variable. $2x$, $7x + 9$, $4ab$, and $\frac{x+5}{2}$ are all algebraic expressions.

Associative properties
Addition: $(a + b) + c = a + (b + c)$
Multiplication: $(ab)c = a(bc)$

Axiom A property assumed to be true without proof. Also called a postulate.

Axis of symmetry A line that divides a parabola into two matching parts.

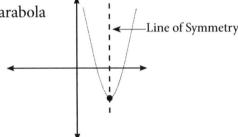

Base The number being multiplied. In an expression such as 4^2, 4 is the base.

Binomial A polynomial with two terms.
$2x + 3y$ and $3x - 2y$ are binomials.

Cartesian coordinate system A system of graphing ordered pairs on a coordinate plane.

Coefficient A number that multiplies the variable. In the term $7x$, 7 is the coefficient of x.

Commutative properties
Addition: $a + b = b + a$
Multiplication: $ab = ba$

Glossary

Completing the square A method for changing a quadratic expression into a perfect square trinomial.

For a quadratic expression in the form $\mathbf{x^2 + bx = c}$, follow these steps:
1) Take half of b which is the coefficient of x
2) Square it.
3) Add the result to both sides of the equal sign.
4) The result will be a perfect square trinomial.

Compound inequality Two or more inequalities that are combined using the word "and" or the word "or."

Constant Specific numbers that do not change.

Coordinates An ordered pair of real numbers, which correspond to a point on a coordinate plane.

Coordinate plane The plane which contains the x and y-axes. It is divided into four quadrants.

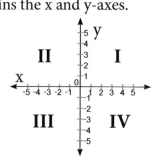

Degree of a polynomial The highest power of a variable that appears in a polynomial expression. The degree of $6x^3 - 4x + 1$ is 3.

Difference of two squares In the difference of two squares, $\mathbf{a^2 - b^2 = (a + b)\,(a - b)}$.

Discriminant The value of $\mathbf{b^2 - 4ac}$ is called the discriminant of the quadratic equation $ax + bx + c = 0$. It allows you to determine which quadratic equations have solutions and which ones do not.

Disjoint sets Sets which have no members in common.
(1, 2, 3) and (4, 5, 6) are disjoint sets.

Distance formula The distance D, between the points (x_1, y_1) and (x_2, y_2) on a coordinate plane is:

$$\mathbf{D \;=\; \sqrt{(x_2 - x_1)^2 + (y_2 - y_1)^2}}$$

Glossary

Distributive Property For real numbers a, b, and c, **a(b + c) = ab + ac.**

Domain of a function The set of all first coordinates (x-values) of the ordered pairs that form the function.

Element of a set Member of a set.

Elimination Method A method of solving a system of linear equations using the following steps:
1) Put the variables on one side of the equal sign and the constant on the other, with the like terms lined up.
2) Add or subtract the equations to eliminate one of the variables. Sometimes it is necessary to multiply one of the equations by a constant first.
3) Solve the equation. Substitute the answer into either of the equations to get the value of the second variable.
4) Check by substituting the answers into the original equations.

Empty set The set that has no members. Also called the null set and is written $\emptyset$ or { }.

Equation A mathematical sentence that contains an equal sign (=) , and states that one expression is equal to another.

Equivalent expressions Expressions which represent the same number.

Evaluate an expression Finding the number an expression stands for by replacing each variable with its numerical value and the simplifying.

Exponent A number that indicates the number of times a given base is used as a factor. In the expression x^2, 2 is the exponent.

Extremes of a proportion In the proportion $\frac{a}{b} = \frac{c}{d}$, a and d are the extremes.

Factor A number or expression that is multiplied to get another number or expression. In the example 4 x 3 = 12, 4 and 3 are factors.

Formula An equation that states a relationship among quantities which are represented by variables. For example, the formula for the area of a rectangle is $A = l \times w$, where A = area, l = length, and w = width.

Function A set of ordered pairs which pairs each x-value with one and only one y-value. For example, F = {(0,2), (-1,6), (4,-2), (-3,4)} is a function.

Glossary

Function notation A way to describe a function that is defined by an equation. In function notation the equation y = 4x – 8 is written as f(x) = 4x – 8, where f(x) is read as "f of x" or "the value of f at x."

Graph To show the points named by numbers or ordered pairs on a number line or coordinate plane.

Greatest common factor The largest factor of two or more numbers or terms. Also written GCF.
The GCF of 15 and 10 is 5, since 5 is the largest number that divides evenly into both 10 and 15.
The GCF of 8ab and 6ab is 2ab.

Grouping symbols Symbols used to group mathematical expressions.
Examples include parentheses ( ), brackets [ ], braces { }, and fraction bars —.

Hypotenuse The side opposite the right angle in a right triangle.

Identity Properties
Addition: **a + 0 = 0 + a**
Multiplication: **1 x a = a**

Inequality A mathematical sentence that states that one expression is greater than or less than another. Inequality symbols are read as follows:
< is less than, ≤ is less than or equal to, > is greater than, ≥ is greater than or equal to

Integers Numbers in the set …-3, -2, -1, 0, 1, 2, 3,… .

Intercept In the equation of a line, the y-intercept is the value of y when x is 0.
The x-intercept is the value of x when y is 0.

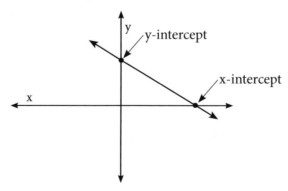

Glossary

Intersection of sets The intersection of two or more sets consists of the members included in all of them. A intersection B is written A∩B.
If set A = {1, 2, 3, 4} and set B = {1, 3, 5} then the intersection sets A and B would be the set {1, 3}.

Inverse operations Operations that "undo" each other. Addition and subtraction are inverse operations. Multiplication and division are inverse operations.

Irrational numbers A real number that cannot be written as the ratio of two integers. They are often represented by non-terminating, non-repeating decimals.
$\sqrt{2} = 1.4142135\ldots$ is an example of an irrational number.

Least Common Denominator (LCD) The least common multiple of the denominators of two or more fractions.
The LCD of $\frac{1}{8}$ and $\frac{1}{6}$ would be 24.
The LCD of $\frac{1}{6x^2}$ and $\frac{1}{4x}$ would be $12x^2$.

Least common multiple (LCM) The least common multiple of two or more expressions is the simplest expression that they will all divide into evenly. To do so, first find the LCM of the coefficients and then the highest degree of each variable and expression.
The LCM of 10x and $25x^2y$ would be $50x^2y$.
The LCM of $4(x-2)^2$ and $(x+2)^2(x-2)$ would be $4(x-2)^2(x+2)^2$.

Like Terms Terms that have the same variables raised to the same power.
$3xy^2$ and $9xy^2$ are like terms. The coefficients do not have to be the same.

Linear Equation An equation that can be written in the form **Ax + By = C**, where A and B are not both zero. The graph of a linear equation is a straight line.

Means of proportion In the proportion $\frac{a}{b} = \frac{c}{d}$, b and c are the means.

Midpoint formula The midpoint between the two points (x_1, y_1) and (x_2, y_2) is
$$M = \left(\frac{x_1 + x_2}{2}, \ \frac{y_1 + y_2}{2} \right)$$

Monomial A term that is a number, a variable, or the product of a number and one or more variables.
5, x, 4xy, 6xy are all examples of monomials.

Glossary

Multiple A multiple of a number is that number multiplied by an integer.

32 is a multiple of 4 since 4 x 8 = 32. Also, 4 and 8 are factors of 32.

Multiplicative inverse Two numbers whose product is one. They are also called reciprocals.

4 and $\frac{1}{4}$ and $-\frac{2}{3}$ and $-\frac{3}{2}$ are examples of multiplicative inverses.

Natural numbers Numbers in the set 1, 2, 3,... . Also called counting numbers.

Negative exponent For any non-zero number x, and any integer n,

$$x^{-n} = \frac{1}{x^n} \text{ and } \frac{1}{x^{-n}} = x^n$$

Negative number A number that is less than zero.

-5 and -3.45 are examples of negative numbers.

Negative slope When the graph of a line slopes down from left to right.

Null set The set that has no members. Also called the empty set which is written Ø and { }.

Number line A line that represents all real numbers with points.

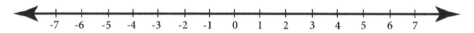

Open sentence A mathematical statement that contains at least one variable.

3x – 6 =12   n > 25   36 = x – 3 are all open sentences.

Ordered pair A pair of numbers (x, y) that represent a point on the coordinate plane. The first number is the x-coordinate and the second is the y-coordinate.

Order of operations The order of steps to be used when simplifying expressions.

1. Inside the grouping symbols.
2. Exponents
3. Multiply and divide in order from left to right.
4. Add and subtract in order from left to right.

Ordinate The second coordinate of an ordered pair. Also called the y-coordinate.

Origin The point where the x-axis and the y-axis intersect in a coordinate plane. Written as (0, 0).

Glossary

Parabola The U-shaped curve that is the graph of a quadratic function.

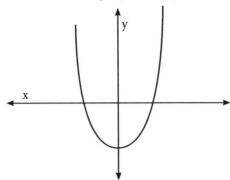

Parallel lines Lines in the same plane that do not intersect. Parallel lines have equal slopes.

Percent Part per hundred or hundredths. Written %.

Perfect square A number than can be expressed as the square of a rational number. The number 36 is a perfect square because it is the square of either 6 or -6.

Perfect square trinomial A trinomial that results from squaring a binomial. Written in the forms $a^2 + 2ab + b = (a + b)^2$ and $a^2 - 2ab + b = (a - b)^2$.

Perpendicular lines Lines in the same plane that intersect at a right (90°) angle. Perpendicular lines have slopes which are negative reciprocals of each other (the product of their slopes is -1).

Point-slope form An equation of a line in the form $y - y_1 = m(x - x_1)$ where m is the slope and (x_1, y_1) is a given point that lies on the line.

Polynomial An algebraic expression of one or more terms connected by plus (+) and minus (–) signs. A monomial has one term. A binomial has two terms. A trinomial has three terms. 3x is a monomial   3 + 5y is a binomial $x^2 + 4x + 3$ is a trinomial

Positive number A number that is greater than 0. 5 and 3.25 are examples of positive numbers.

Positive Slope When the graph of a line slopes up from left to right.

Postulate A property assumed to be true without proof. Also called an axiom

Power An expression that contains a base and an exponent. In the expression x^3, x is the base, and 3 is the exponent.

Prime number A prime number is any whole number greater than 1, whose only factors are one and itself.

Prime factorization A whole number that is expressed as a product of its prime factors. The prime factorization of $100 = 2^2 \times 5^2$.

Proportion An equation that states that two ratios are equal.

Pythagorean theorem In a right triangle, if c is the hypotenuse and a and b are the other two legs, then $\mathbf{a^2 + b^2 = c^2}$.

Quadratic equation An equation that can be written in the form $\mathbf{ax^2 + bx + c = 0}$, where $a \neq 0$.

Quadratic formula The formula that can be used to solve any quadratic equation.

$$x = \frac{-b \pm \sqrt{b^2 - 4ac}}{2a}$$

Quadratic function A function that can be written in the form $\mathbf{y = ax^2 + bx + c}$, where $a \neq 0$.

Radical An expression that is written with a radical sign. The expressions $\sqrt{2}$, $\sqrt{x^2}$, and $\sqrt{25}$ are all radicals.

Radicand The expression that is inside the radical sign.

Range of a function The set of all second coordinates (y-values) of the ordered pairs that form the function.

Ratio A comparison of two numbers using division. Written a:b, a to b, and $\frac{a}{b}$.

Rational Expression A fraction whose numerator and denominator are polynomials. $\frac{x + 3}{x - 2}$ $\frac{x^2 - 6x + 9}{x - 3}$ $\frac{32xy^2}{8xy^2}$ are examples.

Rational numbers A number that can be expressed as the quotient of two integers. The denominator cannot equal zero.

Rationalizing the denominator Changing a fraction that has an irrational denominator to an equivalent fraction that has a rational denominator.

Glossary

Real numbers All positive and negative numbers and zero. This includes fractions and decimals.

Reciprocal The multiplicative inverse of a number. Their product is 1.
The reciprocal of 2 is $\frac{1}{2}$. The reciprocal of $-\frac{2}{3}$ is $-\frac{3}{2}$.

Relation Any set of ordered pairs.

Roots The solutions of a quadratic equation.

Rise The change in **y** going from one point to another on a coordinate plane.
The vertical change.

Run The change in **x** going from one point to another on a coordinate plane.
The horizontal change.

Scientific notation A number written as the product of a number between 1 and 10, and a power of ten. In scientific notation, $7,200 = 7.2 \times 10^3$.

Set A well-defined collection of objects.

Simplified expression The form of an expression with all like terms combined and written in its simplest form.

Slope The steepness of a line. The ratio of the rise (the change in the y direction) to the run (the change in the x direction).
For (x_1, y_1) and (x_2, y_2) which are any two points on a line, **slope** $= \frac{y_2 - y_1}{x_2 - x_1}$, $(x_2 \neq x_1)$

Slope-intercept form An equation of a line in the form $y = mx + b$. The slope is m.
The y-intercept is b.

Solution of an equation A number than can be substituted for the variable in an equation to make the equation true.

Solution of an equation containing two variables An ordered pair (x, y) that makes the equation a true statement.

Square Root If $a^2 = b$, then a is a square root of b. Square roots are written with a radical sign $\sqrt{}$.

Glossary

Standard form of a linear equation $Ax + By = C$, where A and B are not both zero.

Standard form of a quadratic equation $ax^2 + bx + c = 0$, where $a \neq 0$.

Subset If A and B are sets and all the members of set A are members of set B, then set A is a subset of set B.

Substitution method A method of solving a system of linear equations using the following steps:
1. Solve one of the equations for either x or y.
2. Substitute the expression from step 1 into the other equation and solve it for the other variable.
3. Take the value from step 2 and substitute it into either one of the original equations and solve it. You will now have two solutions.
4. Check the solutions in each of the original equations.

System of linear equations Two or more linear equations with the same variables.

System of linear inequalities Two or more linear inequalities in the same variable.

Terms Parts of an expression that are separated by addition or subtraction. Terms can be a number, a variable or a product or quotient of numbers and variables.

7, 3x, 4xy, and -3xy are all examples of terms.

Theorem A statement that can be proven to be true.

Transforming an equation To change an equation into an equivalent equation.

Trinomial A polynomial with three terms.
$x^2 + 2xy + y$ is an example of a trinomial.

Undefined rational expression A rational expression that has zero as a denominator. It is meaningless and is considered undefined.

Union of sets If A and B are sets, the union of A and B is the set whose members are found in set A, or set B, or both set A and set B. A union B is written $A \cup B$.

Variable A letter that represents a number

Variable expression Any expression that contains a variable.

Glossary

Venn diagram A type of diagram which shows how certain sets are related.

Vertex of a parabola The maximum or minimum point of a parabola

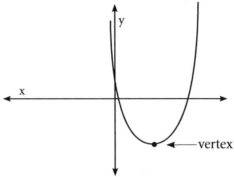

Vertical Line Test A way to tell whether a graph is a function. If a vertical line intersects a graph in more than one point, then the graph is **not** a function.

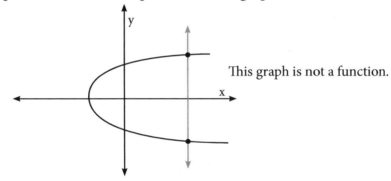

This graph is not a function.

Whole numbers Numbers in the set 0, 1, 2, 3,…

x-coordinate The first number in an ordered pair. Also called the abscissa.

x and y intercepts The points at which a graph intersects the x and y axes.

y-coordinate The second number in an ordered pair. Also called the ordinate.

Zero slope A **horizontal** line has zero slope. The slope of a **vertical** line is undefined.

Zero-product property If the product of two numbers is zero, then at least one of the numbers must equal zero.

Important Formulas

Algebraic Formulas	
Slope formula	$m = \frac{y_2 - y_1}{x_2 - x_1}$
Point-slope form	$y - y_1 = m(x - x_1)$
Slope-intercept form	$y = mx + b$
Standard form of a linear equation	$Ax + By = C$
Quadratic formula	$x = \frac{-b \pm \sqrt{b^2 - 4ac}}{2a}$
Pythagorean theorem	$a^2 + b^2 = c^2$
Distance formula	$D = \sqrt{(x_2 - x_1)^2 + (y_2 - y_1)^2}$
Midpoint formula	$M = \left(\frac{x_1 + x_2}{2}, \frac{y_1 + y_2}{2}\right)$
Other Formulas	
Average speed	$r = \frac{d}{t}$
Interest	$I = p \times r \times t$
Geometric Formulas	
Perimeter of a polygon	P = the sum of the lengths of the sides
Circumference of a circle	$A = \Pi d$
Area of a rectangle	$A = lw$
Area of a square	$A = s^2$
Area of a parallelogram	$A = bh$
Area of a triangle	$A = \frac{1}{2}bh$
Area of a trapezoid	$A = \frac{1}{2}h(b_1 + b_2)$
Area of a circle	$A = \Pi r^2$
Volume of a cube	$V = s^3$
Volume of a rectangular prism	$V = lwh$

Important Symbols

$<$	less than		π	pi
$\leq$	less than or equal to		$\{\ \}$	set
$>$	greater than		$\|\ \|$	absolute value
$\geq$	greater than or equal to		$.\overline{n}$	repeating decimal symbol
$=$	equal to		$1/a$	the reciprocal of a number
$\neq$	not equal to		$\%$	percent
$\cong$	congruent to		(x,y)	ordered pair
$\approx$	is approximately equal to		$f(x)$	f of x, the value of f at x
$(\ )$	parenthesis		$\perp$	perpendicular
$[\ ]$	brackets		$\|\ \|$	parallel to
$\{\ \}$	braces		$\angle$	angle
...	and so on		$\in$	element of
$\bullet$ or $\times$	multiply		$\notin$	not an element of
∞	infinity		$\cap$	intersection
a^n	the nth power of a number		$\cup$	union
$\sqrt{\ }$	square root		$\subset$	subset of
$\varnothing, \{\ \}$	the empty set or null set		$\not\subset$	not a subset of
$\therefore$	therefore		$\triangle$	triangle
$^\circ$	degree			

Multiplication Table

x	2	3	4	5	6	7	8	9	10	11	12
2	4	6	8	10	12	14	16	18	20	22	24
3	6	9	12	15	18	21	24	27	30	33	36
4	8	12	16	20	24	28	32	36	40	44	48
5	10	15	20	25	30	35	40	45	50	55	60
6	12	18	24	30	36	42	48	54	60	66	72
7	14	21	28	35	42	49	56	63	70	77	84
8	16	24	32	40	48	56	64	72	80	88	96
9	18	27	36	45	54	63	72	81	90	99	108
10	20	30	40	50	60	70	80	90	100	110	120
11	22	33	44	55	66	77	88	99	110	121	132
12	24	36	48	60	72	84	96	108	120	132	144

Commonly Used Prime Numbers

2	3	5	7	11	13	17	19	23	29
31	37	41	43	47	53	59	61	67	71
73	79	83	89	97	101	103	107	109	113
127	131	137	139	149	151	157	163	167	173
179	181	191	193	197	199	211	223	227	229
233	239	241	251	257	263	269	271	277	281
283	293	307	311	313	317	331	337	347	349
353	359	367	373	379	383	389	397	401	409
419	421	431	433	439	443	449	547	461	463
467	479	487	491	499	503	509	521	523	541
547	557	563	569	571	577	587	593	599	601
607	613	617	619	631	641	643	647	653	659
661	673	677	683	691	701	709	719	727	733
739	743	751	757	761	769	773	787	797	809
811	821	823	827	829	839	853	857	859	863
877	881	883	887	907	911	919	929	937	941
947	953	967	971	977	983	991	997	1009	1013

Squares and Square Roots

No.	Square	Square Root	No.	Square	Square Root	No.	Square	Square Root
1	1	1.000	51	2,601	7.141	101	10201	10.050
2	4	1.414	52	2,704	7.211	102	10,404	10.100
3	9	1.732	53	2,809	7.280	103	10,609	10.149
4	16	2.000	54	2,916	7.348	104	10,816	10.198
5	25	2.236	55	3,025	7.416	105	11,025	10.247
6	36	2.449	56	3,136	7.483	106	11,236	10.296
7	49	2.646	57	3,249	7.550	107	11,449	10.344
8	64	2.828	58	3,364	7.616	108	11,664	10.392
9	81	3.000	59	3,481	7.681	109	11,881	10.440
10	100	3.162	60	3,600	7.746	110	12,100	10.488
11	121	3.317	61	3,721	7.810	111	12,321	10.536
12	144	3.464	62	3,844	7.874	112	12,544	10.583
13	169	3.606	63	3,969	7.937	113	12,769	10.630
14	196	3.742	64	4,096	8.000	114	12,996	10.677
15	225	3.873	65	4,225	8.062	115	13,225	10.724
16	256	4.000	66	4,356	8.124	116	13,456	10.770
17	289	4.123	67	4,489	8.185	117	13,689	10.817
18	324	4.243	68	4,624	8.246	118	13,924	10.863
19	361	4.359	69	4,761	8.307	119	14,161	10.909
20	400	4.472	70	4,900	8.367	120	14,400	10.954
21	441	4.583	71	5,041	8.426	121	14,641	11.000
22	484	4.690	72	5,184	8.485	122	14,884	11.045
23	529	4.796	73	5,329	8.544	123	15,129	11.091
24	576	4.899	74	5,476	8.602	124	15,376	11.136
25	625	5.000	75	5,625	8.660	125	15,625	11.180
26	676	5.099	76	5,776	8.718	126	15,876	11.225
27	729	5.196	77	5,929	8.775	127	16,129	11.269
28	784	5.292	78	6,084	8.832	128	16,384	11.314
29	841	5.385	79	6,241	8.888	129	16,641	11.358
30	900	5.477	80	6,400	8.944	130	16,900	11.402
31	961	5.568	81	6,561	9.000	131	17,161	11.446
32	1,024	5.657	82	6,724	9.055	132	17,424	11.489
33	1,089	5.745	83	6,889	9.110	133	17,689	11.533
34	1,156	5.831	84	7,056	9.165	134	17,956	11.576
35	1,225	5.916	85	7,225	9.220	135	18,225	11.619
36	1,296	6.000	86	7,396	9.274	136	18,496	11.662
37	1,369	6.083	87	7,569	9.327	137	18,769	11.705
38	1,444	6.164	88	7,744	9.381	138	19,044	11.747
39	1,521	6.245	89	7,921	9.434	139	19,321	11.790
40	1,600	6.325	90	8,100	9.487	140	19,600	11.832
41	1,681	6.403	91	8,281	9.539	141	19,881	11.874
42	1,764	6.481	92	8,464	9.592	142	20,164	11.916
43	1,849	6.557	93	8,649	9.644	143	20,449	11.958
44	1,936	6.633	94	8,836	9.695	144	20,736	12.000
45	2,025	6.708	95	9,025	9.747	145	21,025	12.042
46	2,116	6.782	96	9,216	9.798	146	21,316	12.083
47	2,209	6.856	97	9,409	9.849	147	21,609	12.124
48	2,304	6.928	98	9,604	9.899	148	21,904	12.166
49	2,401	7.000	99	9,801	9.950	149	22,201	12.207
50	2,500	7.071	100	10,000	10.000	150	22,500	12.247

Fraction/Decimal Equivalents

Fraction	Decimal	Fraction	Decimal
$\frac{1}{2}$	0.5	$\frac{5}{10}$	0.5
$\frac{1}{3}$	0.3	$\frac{6}{10}$	0.6
$\frac{2}{3}$	0.6	$\frac{7}{10}$	0.7
$\frac{1}{4}$	0.25	$\frac{8}{10}$	0.8
$\frac{2}{4}$	0.5	$\frac{9}{10}$	0.9
$\frac{3}{4}$	0.75	$\frac{1}{16}$	0.0625
$\frac{1}{5}$	0.2	$\frac{2}{16}$	0.125
$\frac{2}{5}$	0.4	$\frac{3}{16}$	0.1875
$\frac{3}{5}$	0.6	$\frac{4}{16}$	0.25
$\frac{4}{5}$	0.8	$\frac{5}{16}$	0.3125
$\frac{1}{8}$	0.125	$\frac{6}{16}$	0.375
$\frac{2}{8}$	0.25	$\frac{7}{16}$	0.4375
$\frac{3}{8}$	0.375	$\frac{8}{16}$	0.5
$\frac{4}{8}$	0.5	$\frac{9}{16}$	0.5625
$\frac{5}{8}$	0.625	$\frac{10}{16}$	0.625
$\frac{6}{8}$	0.75	$\frac{11}{16}$	0.6875
$\frac{7}{8}$	0.875	$\frac{12}{16}$	0.75
$\frac{1}{10}$	0.1	$\frac{13}{16}$	0.8125
$\frac{2}{10}$	0.2	$\frac{14}{16}$	0.875
$\frac{3}{10}$	0.3	$\frac{15}{16}$	0.9375
$\frac{4}{10}$	0.4		

Solutions

Lesson 1-1
Page 9

1) -7
2) 7
3) -11
4) -101
5) 7
6) -284
7) -160
8) -981
9) -1,540
10) 454
11) -95
12) 15
13) 11
14) -458
15) 500

Lesson 1-2
Page 10

Exercises

1) -16
2) -5
3) -3
4) 11
5) 9
6) -35
7) 49
8) -168
9) 27
10) 6
11) 2
12) -24
13) -202
14) -172
15) 537
16) 106
17) 1
18) -945
19) -100
20) 36

Lesson 1-3
Page 11

Exercises

1) -231
2) -276
3) 152
4) -368
5) 384
6) 441
7) -288
8) -600
9) -312
10) 612
11) -36
12) 72
13) 360
14) -180
15) 72
16) 216
17) 24
18) 168
19) -24
20) -504

Lesson 1-4
Page 12

Exercises

1) -8
2) 16
3) -18
4) -64
5) 170
6) -14
7) -9
8) -23
9) 2
10) 24
11) 24
12) 48
13) 3
14) -30
15) -1
16) 3
17) -1
18) 15
19) 7
20) -2

Lesson 1-5
Page 13

Exercises

1) $\frac{1}{6}$
2) $\frac{7}{20}$
3) $-1\frac{1}{12}$
4) $2\frac{1}{4}$
5) $-\frac{1}{3}$
6) $-\frac{7}{20}$
7) $-\frac{9}{10}$
8) $\frac{1}{6}$
9) $-\frac{3}{20}$
10) $-1\frac{2}{15}$
11) $1\frac{1}{8}$
12) $-1\frac{1}{4}$
13) $-\frac{1}{4}$
14) $2\frac{1}{24}$
15) $-4\frac{1}{2}$
16) 2
17) -7
18) $2\frac{1}{4}$
19) $\frac{1}{10}$
20) $-\frac{11}{12}$

Lesson 1-6
Page 14

Exercises

1) - 2.36
2) -12.69
3) 1.28
4) 5.3562
5) -8.27
6) -1.53
7) -29.28
8) -2.67
9) -1.6
10) -4.338
11) -13
12) 1.344
13) 10.58
14) -3.195
15) -6.384
16) -3.11
17) -27.04
18) 1.05
19) -7.87
20) 12.5

Solutions

Lesson 1-7
Page 15

Exercises

1) 81
2) 9
3) -216
4) 125
5) 2,401
6) 81
7) -512
8) 6,561
9) 256
10) -343
11) 5^5
12) 6^2
13) $(-2)^4$
14) 12^2
15) 15^2
16) $(-9)^3$
17) 10^2
18) 50^2
19) $(-10)^4$
20) 7^2

Lesson 1-8
Page 16

Exercises

1) 6^8
2) $\frac{8^3}{5^3}$
3) $4^3 \times 6^3$
4) 5^{12}
5) 3^5
6) $\frac{1}{6^4}$
7) 7^8
8) $3^3 \cdot 9^3$
9) 4^{10}
10) $\frac{1}{3^3}$
11) 6
12) 1
13) $\frac{1}{4^2}$
14) 5^9
15) $\frac{1}{5^3}$
16) $\frac{7^4}{8^4}$
17) 3^6
18) $\frac{1}{5^3}$
19) 5^9
20) 6^3

Lesson 1-9
Page 17

Exercises

1) 6
2) 11
3) 50
4) 20
5) $5\sqrt[4]{3}$
6) $2\sqrt[4]{2}$
7) 4
8) 60
9) $\frac{\sqrt{5}}{3}$
10) $\frac{2\sqrt{3}}{5}$
11) $5\sqrt[4]{2}$
12) $6\sqrt[4]{2}$
13) $2\sqrt[4]{3}$
14) $10\sqrt[4]{2}$
15) 2
16) $3\sqrt[4]{3}$
17) $5\sqrt[4]{5}$
18) 70
19) 3
20) $\frac{5}{6}$

Lesson 1-10
Page 18

Exercises

1) 31
2) 38
3) 72
4) 48
5) 45
6) -33
7) 83
8) 17
9) 3
10) 2
11) 66
12) 317
13) 6
14) 45
15) 50
16) 2
17) 50
18) -3
19) 29
20) 45

Lesson 1-11
Page 19

Exercises

1) distributive
2) identity of +
3) commutative of +
4) commutative of x
5) inverse of +
6) inverse of x
7) associative of +
8) identity of x
9) inverse of +
10) identity of +
11) commutative of x
12) identity of x
13) distributive
14) commutative of +
15) associative of x
16) inverse of +
17) distributive
18) commutative of x
19) inverse of +
20) commutative of +

Lesson 1-12
Page 20

Exercises

1) -2
2) 10, 1, -3
3) -6, 3, 5
4) 7, 5, 8, -1
5) 9, 7
6) -2, -7, -6, -8
7) 7, 6, 10, -2
8) -6, 5, 4, 7
9) -4, 5, -2, -5
10) -15, 5, 20
11) -5, $\frac{1}{5}$, 15
12) $\frac{3}{5}$, 1, $\frac{7}{5}$
13) $\frac{4}{5}$, $\frac{5}{4}$, $\frac{5}{1}$
14) 2^3, 5^2, 3^3
15) $2\frac{1}{8}$, $3\frac{1}{3}$, $4\frac{1}{2}$
16) $\frac{3}{4}$, 1, $\frac{8}{5}$
17) -103, 100, 180
18) $\frac{1}{4}$, $\frac{1}{2}$, $\frac{3}{4}$
19) $1\frac{7}{8}$, 2, $3\frac{1}{2}$
20) -15, -7, 9

Solutions

Lesson 1-13
Page 21

Exercises

1) (-5, 4)
2) (5, 3)
3) (-3, -5)
4) (1, -2)
5) (0, 4)
6) (-4, 2)
7) (2, -4)
8) (-3, -3)
9) (-2, 3)
10) (5, -4)
11) P
12) M
13) O
14) R
15) J
16) N
17) F
18) Q
19) B
20) L

Lesson 1-14
Page 22

Exercises

1) no - 7 is paired with two y-values
2) yes - each x-value is paired with exactly one y-value
3) 2, 4, 7
4) 3, 5, 8, 9
5) 1, 3, 4
6) 3, 5, 7
7) A vertical line can only intersect a graph at one point. The graph is a function.
8) False
9) True
10) 1, 3, 4, 6
11) 2, 5, 9
12) yes - each x-value is paired with exactly one y-value
13) 3, 5, 7
14) 4, 6, 8, 9
15) no - 5 is paired with two y-values

Lesson 1-15
Page 23

Exercises

1) $2 \times 3 \times 7$
2) $2^2 \times 3^2$
3) 2×3^3
4) $2 \times 5 \times 31$
5) 5^3
6) $2 \times 3 \times 107$
7) 3×31
8) 2×7^2
9) $3 \times 5 \times 7$
10) $2^5 \times 3$
11) $2^4 \times 3^2$
12) $2^3 \times 5^2$
13) $3^3 \times 5$
14) $2^3 \times 3^3$
15) $2^2 \times 3 \times 5$
16) $2^4 \times 3 \times 5$
17) Prime
18) $2 \times 3^2 \times 5$
19) $2 \times 3 \times 17$
20) $2^2 \times 3^2 \times 5$

Lesson 1-16
Page 24

Exercises

1) 5
2) 4
3) 7
4) 6
5) 25
6) 6
7) 2
8) 8
9) 12
10) 11
11) 45
12) 14
13) 3
14) 22
15) 6
16) 105
17) 14
18) 6
19) 21
20) 18

Lesson 1-17
Page 25

Exercises

1) 60
2) 60
3) 66
4) 70
5) 144
6) 210
7) 120
8) 225
9) 196
10) 330
11) 36
12) 180
13) 30
14) 24
15) 60
16) 120
17) 600
18) 240
19) 1,400
20) 720

Solutions

Lesson 1-18
Page 26

Exercises
1) 1.449×10^{10}
2) 3.46×10^{-5}
3) 2.5979×10^{5}
4) 7.21×10^{-6}
5) 1.079×10^{9}
6) 7.6×10^{-3}
7) 1.9×10^{-7}
8) 2.4×10^{5}
9) 7.62×10^{-5}
10) 5.4×10^{7}
11) 609,300
12) 230,000
13) .000006
14) 13,470
15) 321,000
16) .0042
17) 450,000
18) 5,000,000
19) .00372
20) 3,950,000

Lesson 1-19
Page 27

Exercises
1) $3\frac{3}{4}$
2) $1\frac{5}{7}$
3) $3\frac{1}{3}$
4) $\frac{3}{4}$
5) 20
6) 12
7) $13\frac{1}{2}$
8) $3\frac{3}{5}$
9) $6\frac{1}{4}$
10) 32
11) $6\frac{1}{9}$
12) $17\frac{1}{2}$
13) 63
14) $15\frac{3}{4}$
15) 50
16) 2
17) 14
18) 35
19) $2\frac{1}{7}$
20) $6\frac{2}{9}$

Lesson 1-20
Page 28

Exercises
1) $154
2) 5 cans
3) $27^{1}/_{2}$ miles
4) 360 miles
5) 40 liters
6) 80 cups of flour
7) $2.00
8) $2^{1}/_{2}$ cups
9) $1^{3}/_{4}$ inch
10) 35 hours
11) 700 boys
12) $1.65
13) $^{4}/_{15}$ minutes = 16 seconds
14) $4^{1}/_{4}$ inches
15) 25 bags of cement

Lesson 1-21
Page 29

Exercises
1) 42
2) 21
3) 25%
4) 25
5) 20
6) 60
7) 75%
8) 16
9) 25%
10) 20%
11) 5%
12) 25.6
13) 80
14) 24
15) 24
16) 75%
17) 6
18) 320
19) 25%
20) 202.5

Lesson 1-22
Page 30

Exercises
1) 30
2) 30
3) 20%
4) 90%
5) 24 absent
6) 18 caught
7) 20% are red
8) 75% won
9) 250 in the school
10) $25
11) 60%
12) $691.20
13) 75% correct
14) $936
15) 400 enrolled
16) $500
17) 54 correct
18) 75% complete
19) 528 present
20) 50 students

Solutions

Lesson 2-1
Page 31

Exercises

1) n = 21
2) x = 14
3) x = 2
4) x = -23
5) x = 27
6) x = -15
7) x = -16
8) x = 12
9) n = -9
10) n = 54
11) x = 14
12) x = 5.4
13) n = 6
14) m = -6.1
15) x = 1.7
16) x = -6.4

17) n = -22
18) n = -21
19) n = -7
20) n = 1.6

Lesson 2-2
Page 32

Exercises

1) x = 5
2) x = 13
3) x = 63
4) x = -4
5) x = -6
6) x = -15
7) x = 8
8) m = -27
9) m = 60
10) x = -21
11) x = 54
12) x = 15
13) x = 18
14) x = 8
15) x = 7
16) x = 20

17) x = 45
18) x = -45
19) x = 20
20) x = -2

Lesson 2-3
Page 33

Exercises

1) x = 10
2) x = 4
3) x = $7\frac{1}{3}$
4) x = 32
5) x = 12
6) x = 55
7) x = 27
8) x = 240
9) x = 45
10) x = 48
11) x = -49
12) x = 15
13) x = -.4
14) x = 3.4
15) x = -2.4
16) x = 20

17) x = -4
18) x = 4
19) x = 20
20) x = 4

Lesson 2-4
Page 34

Exercises

1) x = 2
2) x = 2
3) x = 7
4) x = 5
5) x = 4
6) x = -13
7) x = 4
8) x = 14
9) x = $-2\frac{4}{5}$
10) x = $6\frac{2}{5}$
11) x = 8
12) x = 1
13) x = $\frac{1}{2}$
14) x = 10
15) x = $10\frac{1}{3}$
16) x = 6

17) x = $2\frac{2}{5}$
18) x = 10
19) x = 1
20) x = 5

Lesson 2-5
Page 35

Exercises

1) x = 2
2) x = 19
3) x = 1
4) x = 17
5) x = 2
6) x = -7
7) x = -20
8) x = -2
9) x = 3
10) x = -17
11) x = -2
12) x = -1
13) x = 0
14) x = $4\frac{1}{2}$
15) x = 9
16) x = -13

17) x = -8
18) x = 23
19) x = -12
20) x = 1

Lesson 2-6
Page 36

Exercises

1) x = 10
2) x = 5
3) x = 19
4) x = 4
5) x = 3
6) x = 2
7) x = 4
8) x = -5
9) x = -8
10) x = -3
11) x = -1
12) x = -5
13) x = 1
14) x = -3
15) x = -14
16) x = 5

17) x = 0
18) x = 5
19) x = 3
20) x = $2\frac{2}{3}$

Solutions

Lesson 2-7
Page 37

Exercises

1) $x = -2$ or $x = 14$
2) $x = -13$ or $x = 7$
3) $x = 10$ or $x = -10$
4) $x = 8$ or $x = -8$
5) $x = 8$ or $x = -5$
6) $x = 5$ or $x = -7\frac{4}{5}$
7) $x = 5$ or $x = -5$
8) $y = 3$ or $y = -5$
9) $x = 3$ or $x = -6$
10) $m = 5$ or $m = -4\frac{3}{5}$
11) $x = 7$ or $x = -11$
12) $x = 7$ or $x = -7\frac{2}{5}$
13) $y = 3$ or $y = -5$
14) $x = 5$ or $x = -1\frac{2}{3}$
15) $x = 8$ or $x = -16$
16) $x = 3$ or $x = 3\frac{2}{3}$
17) $x = 1$ or $x = 3$
18) $x = -11$ or $x = 3$
19) $x = 1$ or $x = 3\frac{2}{3}$
20) $x = 2$ or $x = -8$

Lesson 2-8
Page 38

Exercises

1) $4x + y$
2) $5x - 9y$
3) $15x + 3y$
4) $-6x - 3y$
5) $7y - 16$
6) $5m - 5n - 1$
7) $-15x - 6y + 4$
8) $-7x + 3xy + 2y + 8$
9) $7x + 4y$
10) $-3x - 21y$
11) $-20x - 12y$
12) $x - 2xy + 7y$
13) $16x - 5y + 9z$
14) $-3x^2 - 3y - x^2y$
15) $9x + 4y + 6$
16) $16x^2 - x - xy$
17) $8x^2 - 3xy + 8x - 2$
18) $-2x^2 + 7y^2 - 15$
19) $-3x^2 + 12x - 4y^2 + 14$
20) $9x^2 + 2xy - 5y^2$

Lesson 2-9
Page 39

Exercises

1) $x = -14$
2) $x = -1$
3) $x = -4$
4) $x = 3$
5) $x = 40$
6) $x = 4$
7) $x = 8$
8) $x = 5$
9) $x = -2$
10) $x = 30$
11) $x = -3$
12) $x = 12$
13) $x = 136$
14) $x = -153$
15) $x = 11$
16) $x = 9\frac{2}{7}$
17) $x = -2$
18) $x = 9$
19) $x = -13$
20) $x = 105$

Solutions

Lesson 3-1
Page 40

1)

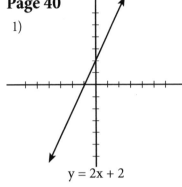

$y = 2x + 2$

2)

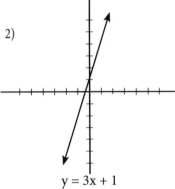

$y = 3x + 1$

3)

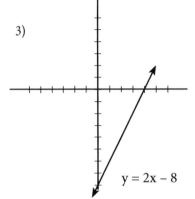

$y = 2x - 8$

4)

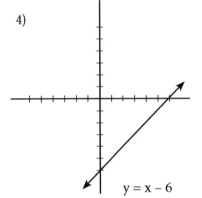

$y = x - 6$

5)

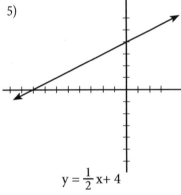

$y = \frac{1}{2}x + 4$

6)

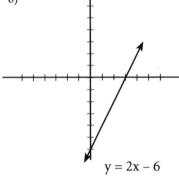

$y = 2x - 6$

7)

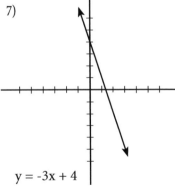

$y = -3x + 4$

8)

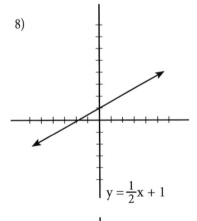

$y = \frac{1}{2}x + 1$

9)

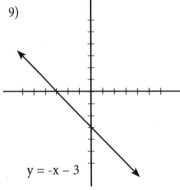

$y = -x - 3$

10)

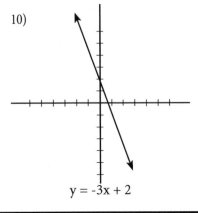

$y = -3x + 2$

11)

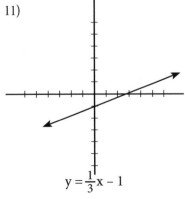

$y = \frac{1}{3}x - 1$

12)

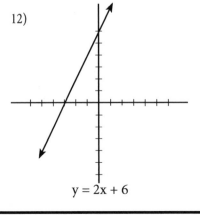

$y = 2x + 6$

Solutions

Lesson 3-1, continued

13)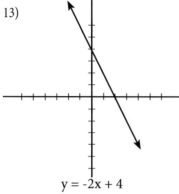

$y = -2x + 4$

14)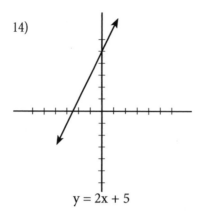

$y = 2x + 5$

15)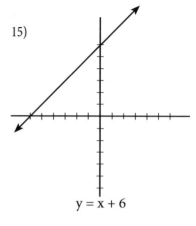

$y = x + 6$

16)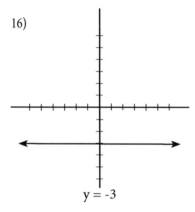

$y = -3$

17)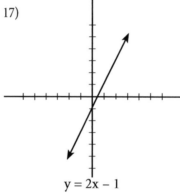

$y = 2x - 1$

18)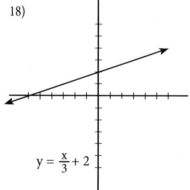

$y = \frac{x}{3} + 2$

19)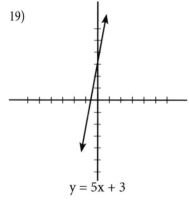

$y = 5x + 3$

20)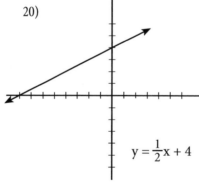

$y = \frac{1}{2}x + 4$

Solutions

Lesson 3-2
Page 41

1)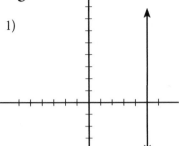
x = 5

x = 5

2)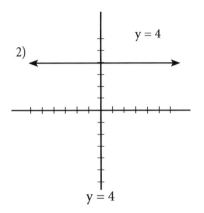
y = 4

y = 4

3)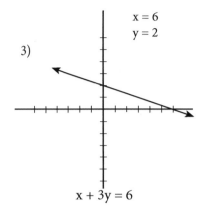
x = 6
y = 2

x + 3y = 6

4)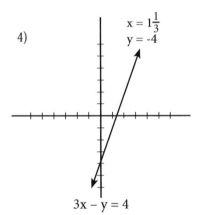
x = 1⅓
y = -4

3x – y = 4

5)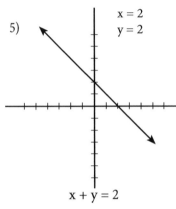
x = 2
y = 2

x + y = 2

6)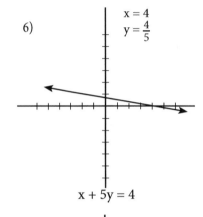
x = 4
y = ⅘

x + 5y = 4

7)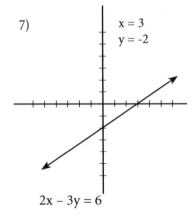
x = 3
y = -2

2x – 3y = 6

8)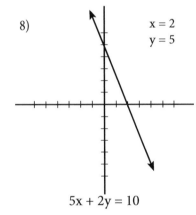
x = 2
y = 5

5x + 2y = 10

9)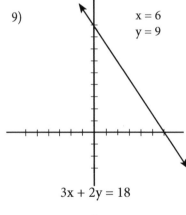
x = 6
y = 9

3x + 2y = 18

10)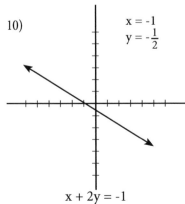
x = -1
y = -½

x + 2y = -1

11)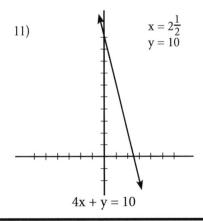
x = 2½
y = 10

4x + y = 10

12)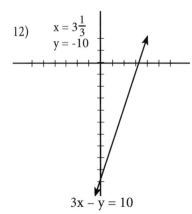
x = 3⅓
y = -10

3x – y = 10

Solutions

Lesson 3-2, continued

13)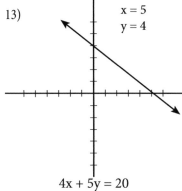
$$x = 5$$
$$y = 4$$
$$4x + 5y = 20$$

14)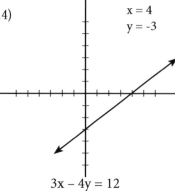
$$x = 4$$
$$y = -3$$
$$3x - 4y = 12$$

15)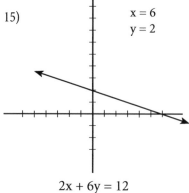
$$x = 6$$
$$y = 2$$
$$2x + 6y = 12$$

16)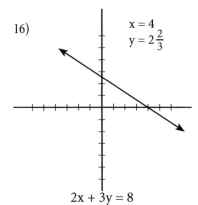
$$x = 4$$
$$y = 2\frac{2}{3}$$
$$2x + 3y = 8$$

17)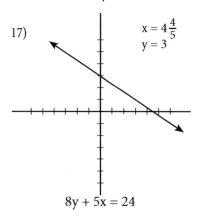
$$x = 4\frac{4}{5}$$
$$y = 3$$
$$8y + 5x = 24$$

18)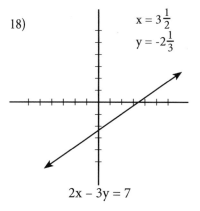
$$x = 3\frac{1}{2}$$
$$y = -2\frac{1}{3}$$
$$2x - 3y = 7$$

19)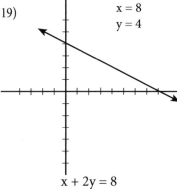
$$x = 8$$
$$y = 4$$
$$x + 2y = 8$$

20)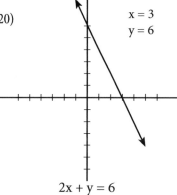
$$x = 3$$
$$y = 6$$
$$2x + y = 6$$

Lesson 3-3
Page 42

Exercises

1) $\frac{3}{2}$

2) $-\frac{1}{2}$

3) -2

4) 1

5) 0

6) $\frac{2}{11}$

7) $\frac{3}{4}$

8) 1

9) undefined

10) $\frac{5}{3}$

11) $\frac{4}{7}$

12) $\frac{2}{3}$

13) 1

14) $\frac{10}{7}$

15) 7

16) $-\frac{1}{2}$

17) -4

18) $-\frac{1}{3}$

19) 3

20) 3

Solutions

Lesson 3-4
Page 43

1)
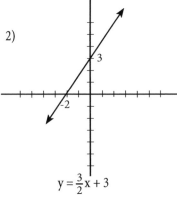
$y = -2x + 5$

2)
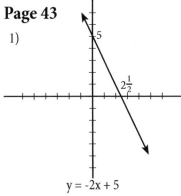
$y = \frac{3}{2}x + 3$

3)
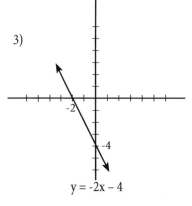
$y = -2x - 4$

4)
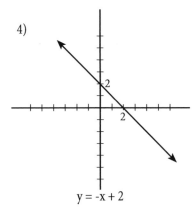
$y = -x + 2$

5)
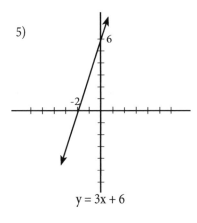
$y = 3x + 6$

6)
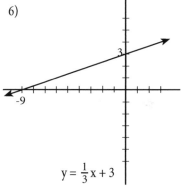
$y = \frac{1}{3}x + 3$

7)
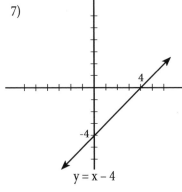
$y = x - 4$

8)
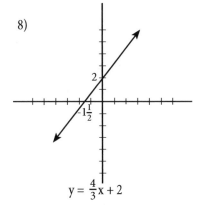
$y = \frac{4}{3}x + 2$

9)
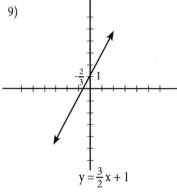
$y = \frac{3}{2}x + 1$

10)
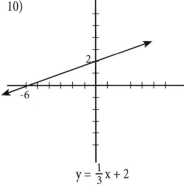
$y = \frac{1}{3}x + 2$

11)
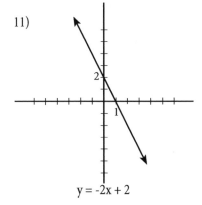
$y = -2x + 2$

12)
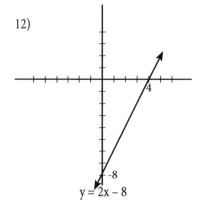
$y = 2x - 8$

Solutions

Lesson 3-4, continued

13)
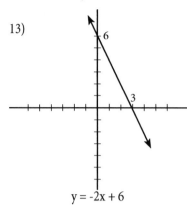

$y = -2x + 6$

14)
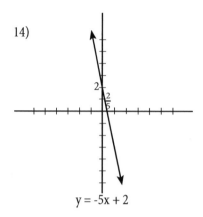

$y = -5x + 2$

15)
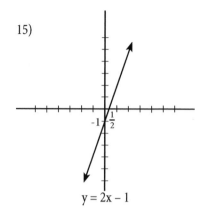

$y = 2x - 1$

16)
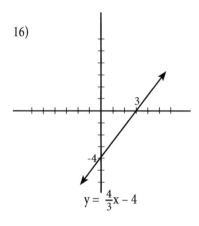

$y = \frac{4}{3}x - 4$

17)
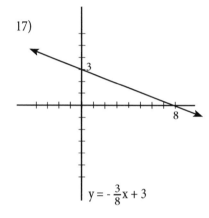

$y = -\frac{3}{8}x + 3$

18)
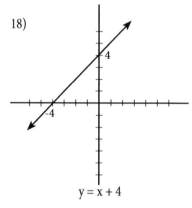

$y = x + 4$

19)
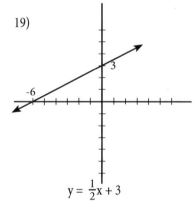

$y = \frac{1}{2}x + 3$

20)
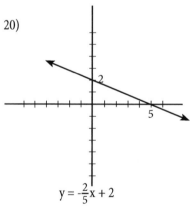

$y = -\frac{2}{5}x + 2$

Solutions

Lesson 3-5
Page 44

1)
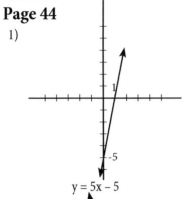
$y = 5x - 5$

2)
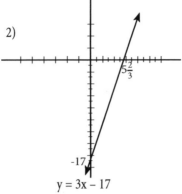
$y = 3x - 17$

3)
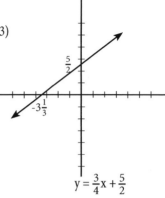
$y = \frac{3}{4}x + \frac{5}{2}$

4)
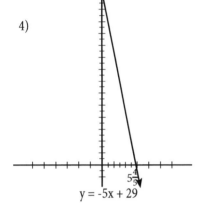
$y = -5x + 29$

5)
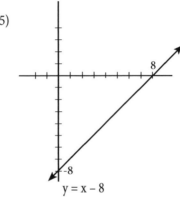
$y = x - 8$

6)
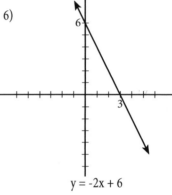
$y = -2x + 6$

7)
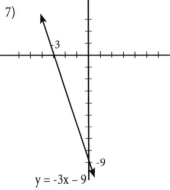
$y = -3x - 9$

8)
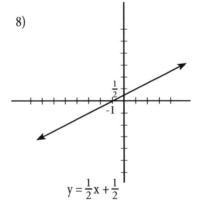
$y = \frac{1}{2}x + \frac{1}{2}$

9)
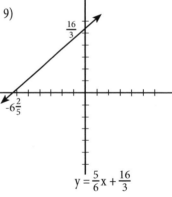
$y = \frac{5}{6}x + \frac{16}{3}$

10)
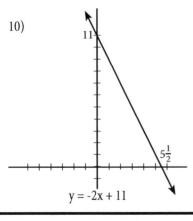
$y = -2x + 11$

11)
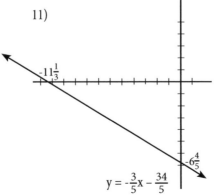
$y = -\frac{3}{5}x - \frac{34}{5}$

12)
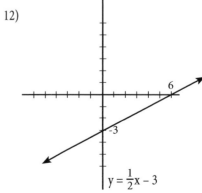
$y = \frac{1}{2}x - 3$

Solutions

Lesson 3-5, continued

13)
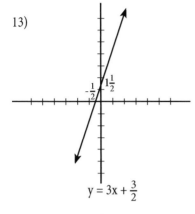
$y = 3x + \frac{3}{2}$

14)
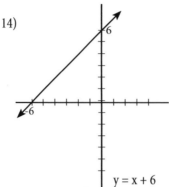
$y = x + 6$

15)
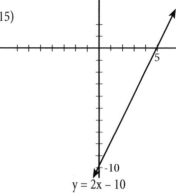
$y = 2x - 10$

16)
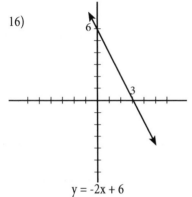
$y = -2x + 6$

17)
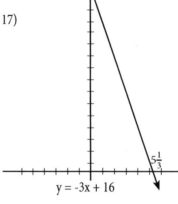
$y = -3x + 16$

18)
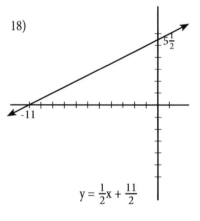
$y = \frac{1}{2}x + \frac{11}{2}$

19)
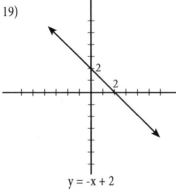
$y = -x + 2$

20)
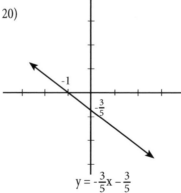
$y = -\frac{3}{5}x - \frac{3}{5}$

Lesson 3-6
Page 45

Exercises

1) $y - 2 = 5(x - 3)$

2) $y - 5 = 3(x - 4)$

3) $y + 5 = 6(x - 3)$

4) $y - 7 = 4(x - 2)$

5) $y - 3 = \frac{3}{5}(x + 2)$

6) $y + 7 = -\frac{2}{3}(x - 4)$

7) $y - 5 = -5(x + 7)$

8) $y + 4 = -\frac{3}{4}(x + 3)$

9) $y - 2 = 5(x - 4)$

10) $y + 6 = -4(x - 4)$

11) $y - 5 = \frac{2}{3}(x - 3)$ or $y - 7 = \frac{2}{3}(x - 6)$

12) $y - 3 = -\frac{5}{8}(x + 4)$ or $y + 2 = -\frac{5}{8}(x - 4)$

13) $y - 1 = -4(x - 5)$ or $y + 3 = -4(x - 6)$

14) $y - 4 = -\frac{2}{3}(x + 6)$ or $y + 2 = -\frac{2}{3}(x - 3)$

15) $y - 5 = \frac{4}{3}(x - 4)$ or $y + 3 = \frac{4}{3}(x + 2)$

16) $y - 2 = \frac{4}{3}(x - 1)$ or $y - 6 = \frac{4}{3}(x - 4)$

17) $y - 4 = 6(x - 2)$

18) $y - 1 = -\frac{1}{3}(x - 5)$

19) $y + 2 = \frac{3}{8}(x - 4)$ or $y + 5 = \frac{3}{8}(x + 4)$

20) $y + 2 = 2(x + 4)$ or $y + 8 = 2(x + 7)$

Solutions

Lesson 4-1
Page 46

Exercises

1)

2)

3)

4)

5)

6)

7)

8)

9)

10)

11)

12)

13)

14)

15)

16)

17)

18)

19)

20)

Lesson 4-2
Page 47

Exercises

1) $x > 8$

2) $x > -4$

3) $x \geq -3$

4) $x \leq 2$

5) $x < -4$

6) $x > -5$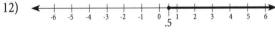

7) $x \leq -4$

8) $x \geq 10$

9) $x \leq 4$

10) $x \geq -14$

11) $x < 6$

12) $x < 6\frac{1}{2}$

13) $x > 3$

14) $x \geq -5$

15) $x > 2\frac{1}{3}$

16) $x < 2$

17) $x > 1$

18) $x \leq -5$

19) $x < 1$

20) $x \leq -6$

Solutions

Exercises

1) $x \geq -8$ ————•————————→ -8

2) $x \geq 5$ ←————•———→ 5

3) $x \geq 48$ ←————————•→ 48

4) $x > -30$ ←○————————→ -30

5) $x > -3$ ←————○———→ -3

6) $x > -8$ ←——○—————→ -8

7) $x > -28$ ←○————————→ -28

8) $x < -3$ ←————○———→ -3

9) $x \geq -14$ ←—•————————→ -14

10) $x > 7$ ←————————○——→ 7

11) $x > 3$ ←——————○——→ 3

12) $x < 21$ ←————————————○→ 21

13) $x > 7$ ←————————○——→ 7

14) $x \geq 8$ ←——————————•→ 8

15) $x \geq 5$ ←——————•———→ 5

16) $x > -4$ ←———○————→ -4

17) $x \leq 6$ ←——————•———→ 6

18) $x \geq 12$ ←——————————•——→ 12

19) $x > -18$ ←○————————→ -18

20) $x \leq 14$ ←————————————•—→ 14

Exercises

1) $x > 8$ ←——————————○—→ 8

2) $x < 8$ ←——————————○—→ 8

3) $x < -3$ ←———○————→ -3

4) $x \leq 9$ ←——————————•—→ 9

5) $x > -25$ ←○————————→ -25

6) $m < 2$ ←————————○——→ 2

7) $x \geq 2\frac{1}{2}$ ←————————•———→ $2\frac{1}{2}$

8) $x > -2$ ←————○—————→ -2

9) $x > 56$ ←————————————○→ 56

10) $x \geq 19$ ←——————————•—→ 19

11) $x < 4\frac{1}{16}$ ←————————○———→ $4\frac{1}{16}$

12) $x \geq -19\frac{1}{2}$ ←•————————————→ $-19\frac{1}{2}$

13) $x \leq 11$ ←————————————•——→ 11

14) $x > -16$ ←○————————→ -16

15) $x \geq 3$ ←——————————•—→ 3

16) $x > -4$ ←———○————→ -4

17) $x > -2$ ←————○—————→ -2

18) $x \geq 4$ ←——————•————→ 4

19) $x < -2$ ←————○—————→ -2

20) $x > 9$ ←——————————○—→ 9

Solutions

Lesson 4-5
Page 50

Exercises

1) $-4 < x < 2$

2) $-1 < x < 3$

3) $x \le 11$ or $x \ge 16$

4) $-5 \le x < 4$

5) $x \le -\frac{3}{2}$ or $x > 4$

6) $3 \le x < 12$

7) $1 < x < 3$

8) $x < 2$ or $x > 7$

9) $x \le -1$ or $x > 9$

10) $x < 4$ or $x > 5$

11) $-1 \le x < 4$

12) $-6 < x < 1$

13) $1 < x < 6$

14) $x < 5$ or $x > 11$

15) $0 < x \le 3$

16) $0 < x \le 4$

17) $-5 \le x < 0$

18) $-5 < x < 4$

19) $-5 \le x < 13$

20) $x < -1$ or $x \ge 2$

Lesson 4-6
Page 51

Exercises

1) $x < -8$ or $x > 8$

2) $-5 \le x \le 5$

3) $-5 < x < 5$

4) $x < -14$ or $x > 4$

5) $-3 \le x \le 7$

6) $x \ge 7$ or $x \le -1$

7) $0 \le x \le 12$

8) $-14 < x < 4$

9) $x \ge -1$ or $x \le -7$

10) $x \le 1$ or $x \ge \frac{13}{3}$

11) $-1 < x < 7$

12) $x < -1$ or $x > 7$

13) $-4 < x < 4$

14) $-3 \le x \le 2$

15) $x \le -6$ or $x \ge 6$

16) $-6 \le x \le 3$

17) $-1 < x < 1$

18) $x \le -9$ or $x \ge 9$

19) $x \le -8$ or $x \ge 7$

20) $-8 \le x \le -2$

Solutions

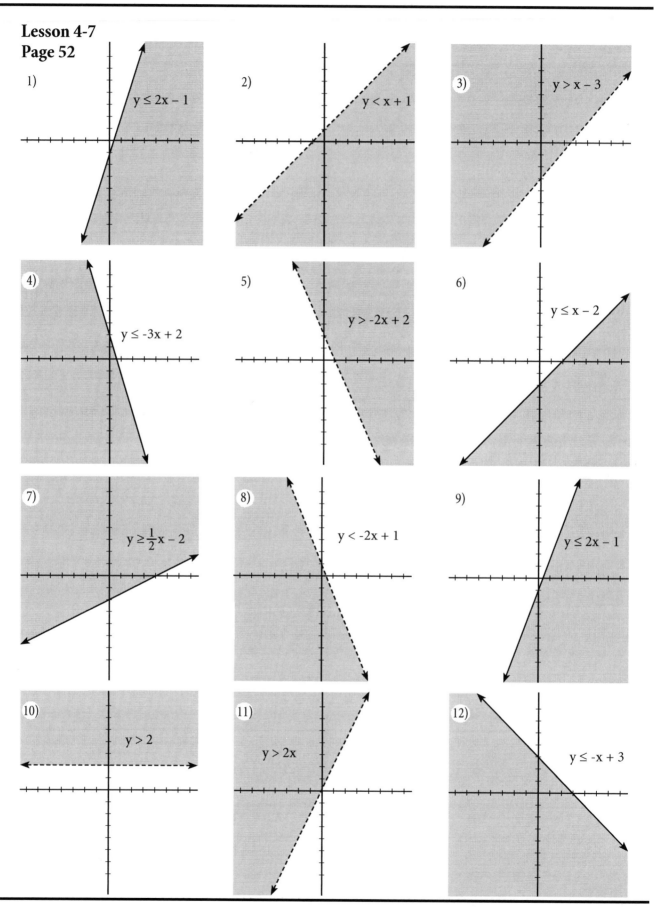

1) $y \leq 2x - 1$

2) $y < x + 1$

3) $y > x - 3$

4) $y \leq -3x + 2$

5) $y > -2x + 2$

6) $y \leq x - 2$

7) $y \geq \frac{1}{2}x - 2$

8) $y < -2x + 1$

9) $y \leq 2x - 1$

10) $y > 2$

11) $y > 2x$

12) $y \leq -x + 3$

Solutions

Lesson 4-7, continued

13) 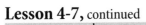 $y \geq \frac{1}{3}x + 2$

14) $y < 2x - 6$

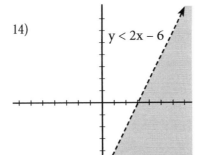

15) $x < 2$

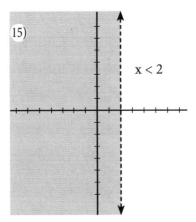

16) 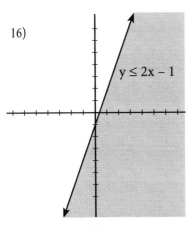 $y \leq 2x - 1$

17) $y < -x + 3$

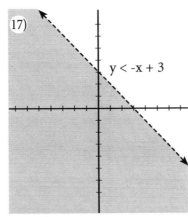

18) $y > 3x + 4$

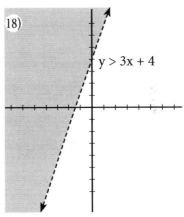

19) $y > x - 1$

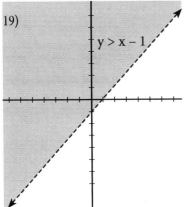

20) $y \leq -2x + 4$

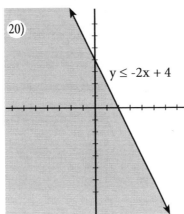

Solutions

Lesson 5-1
Page 53

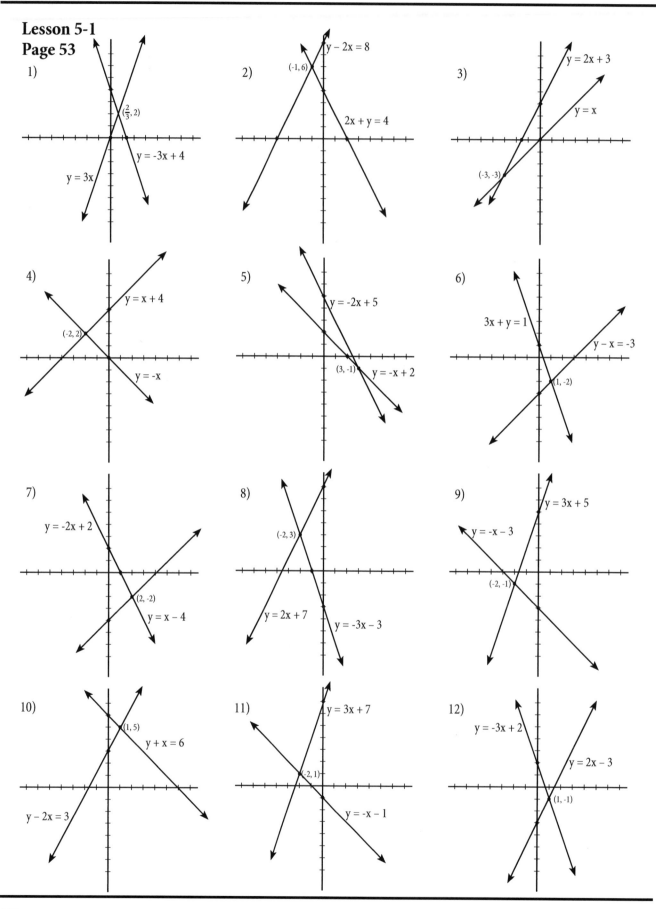

1)

$(\frac{2}{3}, 2)$

$y = -3x + 4$

$y = 3x$

2)

$y - 2x = 8$

$(-1, 6)$

$2x + y = 4$

3)

$y = 2x + 3$

$y = x$

$(-3, -3)$

4)

$y = x + 4$

$(-2, 2)$

$y = -x$

5)

$y = -2x + 5$

$(3, -1)$ $y = -x + 2$

6)

$3x + y = 1$

$y - x = -3$

$(1, -2)$

7)

$y = -2x + 2$

$(2, -2)$

$y = x - 4$

8)

$(-2, 3)$

$y = 2x + 7$

$y = -3x - 3$

9)

$y = 3x + 5$

$y = -x - 3$

$(-2, -1)$

10)

$(1, 5)$

$y + x = 6$

$y - 2x = 3$

11)

$y = 3x + 7$

$(-2, 1)$

$y = -x - 1$

12)

$y = -3x + 2$

$y = 2x - 3$

$(1, -1)$

Solutions

Lesson 5-1, continued

13)

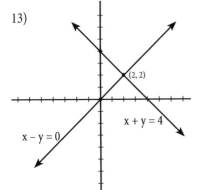

$x - y = 0$

$x + y = 4$

$(2, 2)$

14)

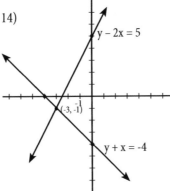

$y - 2x = 5$

$(-3, -1)$

$y + x = -4$

15)

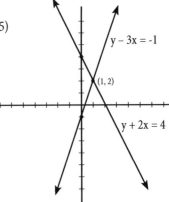

$y - 3x = -1$

$(1, 2)$

$y + 2x = 4$

16)

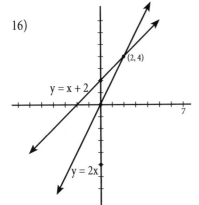

$(2, 4)$

$y = x + 2$

$y = 2x$

17)

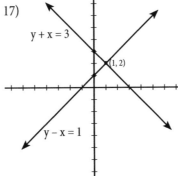

$y + x = 3$

$(1, 2)$

$y - x = 1$

18)

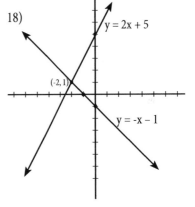

$y = 2x + 5$

$(-2, 1)$

$y = -x - 1$

19)

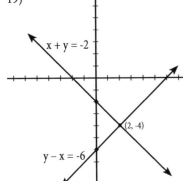

$x + y = -2$

$(2, -4)$

$y - x = -6$

20)

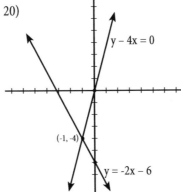

$y - 4x = 0$

$(-1, -4)$

$y = -2x - 6$

Solutions

Lesson 5-2
Page 54

Exercises

1) (4, 2)
2) (2, 10)
3) (1, 2)
4) (1, 3)
5) (3, 9)
6) (-2, 8)
7) (1, 3)
8) (2, 10)
9) (1, 2)
10) (-5, 25)
11) (1, 3)
12) (6, 18)
13) (4, 24)
14) (7, 5)
15) (2, 7)
16) (1, 4)
17) (-7, 22)
18) (5, -3)
19) (12, 4)
20) (-3, -5)

Lesson 5-3
Page 55

Exercises

1) (9, 1)
2) (1, -1)
3) (3, 5)
4) (4, 4)
5) (3, 0)
6) (3, 4)
7) $(-\frac{1}{2}, 3)$
8) (-5, -2)
9) (-3, -5)
10) (3, -2)
11) (4, 5)
12) (2, -2)
13) (4, 1)
14) (-1, 3)
15) (3, 2)
16) (3, -1)
17) $(2, -\frac{1}{3})$
18) (-2, 3)
19) (-1, 3)
20) (-3, -12)

Lesson 5-4
Page 56

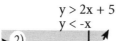

1)
$$y > 2x - 1$$
$$y < -x + 3$$

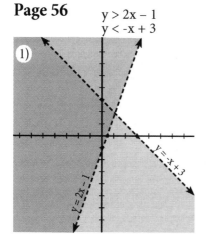

2)
$$y > 2x + 5$$
$$y < -x$$

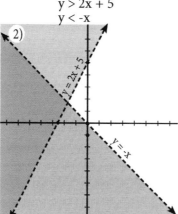

3)
$$y > x$$
$$y > -2x + 4$$

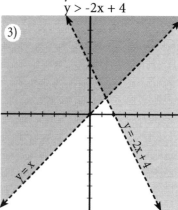

4)
$$y \leq -2x + 4$$
$$y \geq 3x + 6$$

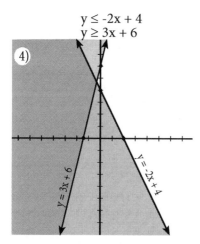

5)
$$y > -x - 1$$
$$y \leq 2x + 1$$

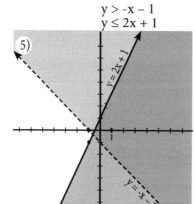

6)
$$y > 2x$$
$$y < -2x + 4$$

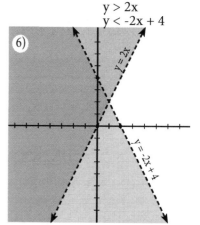

Solutions

Lesson 5-4, continued

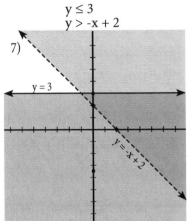

7) $y \leq 3$
$y > -x + 2$

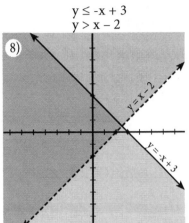

8) $y \leq -x + 3$
$y > x - 2$

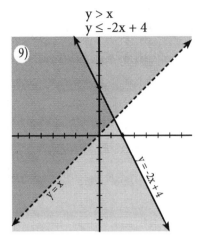

9) $y > x$
$y \leq -2x + 4$

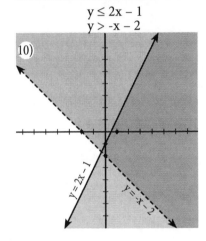

10) $y \leq 2x - 1$
$y > -x - 2$

Lesson 6-1
Page 57

Exercises
1) $6x^2 + 6x + 11$
2) $6x^2 + 5x - 5$
3) $3x + 8$
4) $3x + 1$
5) $5x^4 - x^3 - 15$
6) $-2n^2 + 7n + 5$
7) $2y^2 - 6y + 8$
8) $6x^2 - 7x + 15$
9) $4x^2 + 2x - 3$
10) $15x^2 + 29x + 8$
11) $x^2 - 7x - 11$
12) $4x^2 - 5xy + 6y^2$
13) $21x^2 - 12x - 5$
14) $4x^2 + 6x - 2$
15) $10x - 13$
16) $-x^2 - 2xy + 2y^2$
17) $7x^2 - xy + 4y^2$
18) $2x^2 + 4$
19) $5x^2 - 4x + 3$
20) $5x + 8y + z$

Lesson 6-2
Page 58

Exercises
1) $2x - 9y$
2) $-4x - 2$
3) $-x^2 - 6x - 5$
4) $-x^2 + 5x - 1$
5) $3x^2 - 9xy + 8y^2$
6) $5x + 5$
7) $-7x - 11y + 12$
8) $2x^3 - 3y^2 - 6$
9) $4x^2y^2 - 15xy + 34y^2$
10) $2x - 9y$
11) $x + 2$
12) $-x^2 + 4$
13) $5x^2 + x + 3$
14) $x^2 - xy + 6$
15) $-4x^2 - 6xy - 13$
16) $-8y^2 - 8$
17) $5x^3 - 3x^2 - 2$
18) $-2x^3 - 11$
19) $-3x + 3$
20) $-9x^2 + 2x + 5$

Solutions

Lesson 6-3
Page 59

Exercises
1) x^{12}
2) $5x^7$
3) $-81x^6$
4) $4xy^5$
5) $-49x^8y^5$
6) $18x^7y^8$
7) $27x^6$
8) $25x^4$
9) $-4x^4y^4z^3$
10) $125x^9$
11) $24x^5y^5$
12) $81x^2y^2$
13) $81x^4$
14) $-36x^3y^4$
15) $-20x^{10}y^7$
16) $-24x^3y^6$
17) $-27x^9$
18) $144x^2y^6$
19) $15x^9$
20) $-27x^4y^5$

Lesson 6-4
Page 60

Exercises
1) $7x^2$
2) $-9x^2$
3) $8x^2y^4$
4) xy^2
5) $\frac{-15x}{y}$
6) $\frac{10y^2}{x}$
7) $-4x$
8) $\frac{6xy}{z}$
9) $\frac{1}{xyz}$
10) $3x^2$
11) $3x^3$
12) y^2
13) $\frac{3x^2}{y}$
14) $\frac{3x}{y}$
15) $-4x^2y^2$
16) $\frac{-9xy^3}{z^2}$
17) $\frac{6x^4}{y^6}$
18) $\frac{-3x^2}{y}$
19) $4x$
20) $\frac{-5x^4}{y^7}$

Lesson 6-5
Page 61

Exercises
1) $8x^2 - 12x$
2) $12x^3 + 24x^2$
3) $15x^6 + 20x^3$
4) $15x^4 + 30x^3 - 35x^2$
5) $-24x^6 + 12x^4 - 16x^2$
6) $6y^4 - 12y^3z + 3y^2z^3$
7) $21x^3 + 14x^2y^2 - 21x^3y$
8) $8x^3y^3 - 4x^3y^2 + 6x^4y^3$
9) $20x^2y^2 - 30x^3y^2 + 10x^3y$
10) $-20x^3 - 12x^4y - 4x^2y$
11) $18x^2y + 24y^4$
12) $-6x^4 + 8x^3 - 18x^2$
13) $40x^3y + 16x^2y^3 - 24x^2y$
14) $-66x^4 + 18x^3 + 30x^2$
15) $-10x^2y^2 - 25x^3y^2 + 5x^2y^3 - 5x^5y$
16) $30a^3b - 10ab^2 + 15a^3b^3$
17) $15x^2 + 10xy - 15x^2y^3$
18) $7x^4y^2 + 2x^5y^2 - 7x^3$
19) $8x^3 + 6x^2 - 10x$
20) $12x^4y^2 + 4x^2y^4 + 36x^2y^5 + 12x^3y^2$

Lesson 6-6
Page 62

Exercises
1) $x^2 + 10x + 16$
2) $x^2 + 13x + 42$
3) $x^2 - 13x + 36$
4) $x^2 - x - 6$
5) $5x^2 + 4x - 12$
6) $x^2 - 8x + 15$
7) $2x^2 + 9x - 35$
8) $3x^2 + 34x - 24$
9) $x^2 - 4x - 32$
10) $40x^2 + 31x + 6$
11) $-24x^2 - 28x - 8$
12) $18x^2 - 63x + 40$
13) $4x^2 + 13x - 12$
14) $2x^2 - 41x + 155$
15) $4x^2 + 20xy + 25y^2$
16) $6x^2 - 11xy - 7y^2$
17) $10x^2 - 33x + 27$
18) $x^2 - 18x + 81$
19) $7x^2 - 30x + 27$
20) $12x^2 - 28xy + 15y^2$

Lesson 6-7
Page 63

Exercises
1) $4x + 9$
2) $1 - 2x - x^4$
3) $5x^2 + 8x - 2$
4) $5x^2 + 3x - 6$
5) $-4x^4 + 4x^2 + 1$
6) $-3xy - x + 2y$
7) $y^2 - 3y + 5$
8) $x^2y^2 - xy + 1$
9) $-3x^2y^2 - 4y + 1$
10) $3x + 6y$
11) $xy - 3y^2$
12) $-9x^3 - 4y^3 + 3y^2$
13) $-7x^4 - 8x^2 + 9$
14) $-5x^2 + 7x - 8\frac{3}{4}$
15) $3x + 3 - 3x^2$
16) $-4x^2y^3 + 3xy + 1$
17) $x - 3y$
18) $3x + 4y + 5$
19) $2y - 1 + 3x$
20) $7xy - 5$

Lesson 6-8
Page 64

Exercises
1) $x + 4$
2) $x + 8$
3) $x - 7$
4) $x + 2$
5) $3x + 5$
6) $3x + 4$
7) $x - 2$
8) $7x - 1$
9) $x - 3$
10) $3x - 8$
11) $6x + 5$
12) $2x + 1 + \frac{2}{4x + 1}$
13) $3x + 8$
14) $n + 6$
15) $x - 3 + \frac{8}{x + 1}$
16) $x + 4 - \frac{1}{3x - 2}$
17) $x - \frac{9}{x - 3}$
18) $2x - 5$
19) $x - 18$
20) $x + 5$

Solutions

Lesson 6-9
Page 65

Exercises
1) $3x(x-1)$
2) $2x(x+3)$
3) $7x^2(4+3x^2)$
4) $2x(7x^2-1)$
5) $3(4x^2+3x+2)$
6) $2xy^2(x^2+3y+4)$
7) $16x(x^5-2x^4-3)$
8) $5xy^2(2x^2-3)$
9) $8xy(y+3)$
10) $x(1+xy^2+x^2y^3)$
11) $x(15xy^2+25y+1)$
12) $x^2y^2(6xy+3x+1)$
13) $2x^3(x^4-x^3-8x^2+2)$
14) $10x(x+4)$
15) $5xy(3-7y)$
16) $5xy^2(5x+6y)$
17) $8x(2x+1)$
18) $2xy(x^2y-8y+4)$
19) $9y(3x^2+y^2)$
20) $5x(y+3x)(y-x)$

Lesson 6-10
Page 66

Exercises
1) $(x+3)(x+5)$
2) $(x+3)(x+9)$
3) $(x+10)(x+3)$
4) $(x+2)(x+4)$
5) $(x-7)(x-2)$
6) $(x+5)(x-3)$
7) $(x-8)(x+3)$
8) $(x-12)(x+3)$
9) $(x+3)(x+12)$
10) $(x+3)(x-1)$
11) $(x+8)(x-5)$
12) $(x-1)(x-7)$
13) $(x+5)(x+7)$
14) $(x-4)(x-15)$
15) $(x-5)(x-2)$
16) $(x+7)(x-1)$
17) $(x-7)(x+6)$
18) $(x+12)(x-6)$
19) $(x+4)(x+3)$
20) $(x-6)(x+9)$

Lesson 6-11
Page 67

Exercises
1) $(x+6)(x-6)$
2) $(10-x)(10+x)$
3) $(4n+3)(4n-3)$
4) $(2x+5)(2x-5)$
5) $(5m+7)(5m-7)$
6) $25(2x+1)(2x-1)$
7) $(x+3)(x-3)$
8) $(12x+7y)(12x-7y)$
9) $(10x+y)(10x-y)$
10) $(5+2y)(5-2y)$
11) $(1+3x)(1-3x)$
12) $(7-xy)(7+xy)$
13) $(5mn+4xy)(5mn-4xy)$
14) $(4m+3n)(4m-3n)$
15) $(4y+5)(4y-5)$
16) $(9y^3+5y)(9y^3-5y)$
17) $(x^2+4)(x+2)(x-2)$
18) $(8x+5y)(8x-5y)$
19) $(9x^2+1)(3x+1)(3x-1)$
20) $(x^4+y^3)(x^4-y^3)$

Lesson 6-12
Page 68

Exercises
1) $2(a+b)(a-b)$
2) $3(x+3)(x-3)$
3) $3(x+2)(x-4)$
4) $(x+4)(x-4)$
5) $2(x-5)(x-3)$
6) $5(x+2)(x-6)$
7) $2(x-4)^2$ or $2(x-4)(x-4)$
8) $x(x+2)(x-2)$
9) $6(x+2y)(x-2y)$
10) $2(x+9)(x-9)$
11) $3(x+1)^2$ or $3(x+1)(x+1)$
12) $2(2x+1)(x-2)$
13) $x(2x+7)(2x-7)$
14) $x^2(4+y^2)(2+y)(2-y)$
15) $10x(x+2)(x-2)$
16) $(t^2+4)(t+2)(t-2)$
17) $3(x^2+1)(x+1)(x-1)$
18) $2(x+8)(x-8)$
19) $3(x+5)(x-5)$
20) $7(x+1)(x-1)$

Lesson 6-13
Page 69

Exercises
1) $(3x+1)(x+3)$
2) $(2x+3)(x-1)$
3) $(3x+1)(x-2)$
4) $(3x-5)(x+1)$
5) $(2x-5)(2x-1)$
6) $(3x+4)(2x-1)$
7) $(2x+1)(x-4)$
8) $(5x-9)(x+2)$
9) $(3x+1)(x+1)$
10) $(2x+1)(x-1)$
11) $(3x+1)(x-2)$
12) $(3x+1)(x+2)$
13) $(3x-5)(x-1)$
14) $(5x-1)(x+1)$
15) $(7x+1)(x+3)$
16) $(2x+5)(x+1)$
17) $(3x+4)(3x+2)$
18) $(3x+2)(x+1)$
19) $(2x+3)(x-6)$
20) $(2x+1)(x+2)$

Solutions

Lesson 7-1
Page 70

Exercises

1) $6n^2$

2) $\dfrac{6x}{y}$

3) $25xy$

4) $\dfrac{5x^4 y^2}{z}$

5) $16x^2 y$

6) $\dfrac{6x^2}{y}$

7) $-4x^5 y$

8) $\dfrac{3x}{yz}$

9) $\dfrac{5}{xy}$

10) $\dfrac{7xz^2}{y}$

11) $\dfrac{8}{xy^2}$

12) $\dfrac{1}{x^2 yz}$

13) $\dfrac{4y^3}{x^2}$

14) $\dfrac{5}{xy^3}$

15) $\dfrac{x}{y}$

16) $\dfrac{3}{xz^2}$

17) $\dfrac{4x}{yz}$

18) $\dfrac{27x^3}{yz}$

19) $\dfrac{5xy}{z}$

20) $\dfrac{5}{x}$

Lesson 7-2
Page 71

Exercises

1) $\dfrac{2}{3}$

2) $\dfrac{3}{x+1}$

3) $\dfrac{3(x-2)}{(x-4)}$

4) $\dfrac{2(y-1)}{9}$

5) $x-4$

6) $x+8$

7) $\dfrac{4}{x+4}$

8) $\dfrac{2}{x+5}$

9) $\dfrac{x+6}{x+1}$

10) $\dfrac{x+3}{x+9}$

11) $\dfrac{x+2}{x-2}$

12) $\dfrac{x-5}{x-4}$

13) $\dfrac{3}{x-2}$

14) x

15) $\dfrac{x(x-2)}{x-3}$

16) $\dfrac{x-5}{x-2}$

17) $\dfrac{6}{x-3}$

18) $\dfrac{x-3}{x+3}$

19) $\dfrac{x-3}{x+2}$

20) $\dfrac{3}{x-1}$

Lesson 7-3
Page 72

Exercises

1) -1

2) -1

3) $-\dfrac{1}{2}$

4) $-(x+5) = -x-5$

5) $-(x+1) = -x-1$

6) $-\dfrac{1}{x+4}$

7) $-\dfrac{1}{x+2}$

8) -1

9) $-x-5$

10) $-\dfrac{1}{x+7}$

11) $-(m+n) = -m-n$

12) -1

13) -1

14) $-(m+2) = -m-2$

15) $\dfrac{x+y}{-5}$

16) $-\dfrac{1}{x+6}$

17) -1

18) $\dfrac{-2(2x+y)}{x+y}$

19) -1

20) $-\dfrac{1}{3}$

Lesson 7-4
Page 73

Exercises

1) $x = 27$

2) $x = 15$

3) $x = 8$

4) $x = 5$

5) $x = 18$

6) $x = 8$

7) $x = 7$

8) $x = 4$

9) $x = 8$

10) $x = -10$

11) $y = 2$

12) $m = -7$

13) $x = 3$

14) $x = 2$

15) $x = 5$

16) $x = 2$

17) $x = 3$

18) $x = 11$

19) $x = 5$

20) $x = 4$

Lesson 7-5
Page 74

Exercises

1) 3

2) 12

3) $2x$

4) $\dfrac{2r}{n}$

5) $\dfrac{2b^2}{3a}$

6) $\dfrac{r}{2}$

7) $\dfrac{2}{5}$

8) $\dfrac{1}{5(x+4)}$

9) $\dfrac{(x+5)^2}{9}$

10) $\dfrac{y-2}{y-1}$

11) $\dfrac{x}{x+3}$

12) $\dfrac{2(y-3)}{5}$

13) $\dfrac{3xy}{10}$

14) $8y$

15) $\dfrac{x+3}{5}$

16) $2(x+2)$

17) n

18) $\dfrac{x+y}{2}$

19) $2y$

20) $\dfrac{n-4}{n+4}$

Solutions

Lesson 7-6
Page 75

Exercises

1) $\frac{1}{3}$

2) $\frac{4}{7}$

3) $4xy$

4) $\frac{3}{x-1}$

5) $\frac{ab}{4c}$

6) $\frac{3(x+5)}{2}$

7) $\frac{3m^3}{5}$

8) $\frac{15}{8} = 1\frac{7}{8}$

9) $\frac{a-5}{3(a-1)}$

10) $\frac{b+3}{4b}$

11) $\frac{27x}{2y^2}$

12) $8(x-1)$

13) $\frac{8}{x(x-y)}$

14) $\frac{2}{y^2}$

15) $\frac{3}{4(x-5)}$

16) $\frac{y-4}{y+8}$

17) $2x$

18) $\frac{x(x-4)}{3}$

19) $\frac{4d}{9c}$

20) $\frac{6x}{x-5}$

Lesson 7-7
Page 76

Exercises

1) x

2) $\frac{2x}{7}$

3) $\frac{11}{2x}$

4) $\frac{5x-7}{7}$

5) 2

6) 1

7) $\frac{x-3}{x+3}$

8) $\frac{5x-2}{x-2}$

9) $\frac{3x-5}{3}$

10) $y-6$

11) $\frac{5a-1}{a-1}$

12) $\frac{1}{3}$

13) $\frac{4b-23}{2b+12}$

14) 2

15) $\frac{3x+1}{x-4}$

16) $\frac{a+b}{ab}$

17) $\frac{7(x+1)}{x+2}$

18) $\frac{-y+4}{y+5}$

19) $\frac{1}{x+3}$

20) $x+y$

Lesson 7-8
Page 77

Exercises

1) $\frac{7x}{6}$

2) $\frac{7x}{6}$

3) $\frac{3+x}{2x}$

4) $\frac{y-10}{5y}$

5) $\frac{4+5b}{2ab}$

6) $\frac{7x^2}{8}$

7) $\frac{4y+1}{6xy}$

8) $\frac{7x}{24}$

9) $\frac{47}{12x}$

10) $-\frac{x}{21}$

11) $\frac{22}{15x}$

12) $\frac{9m-7}{18}$

13) $\frac{7}{12x}$

14) $\frac{1}{12x}$

15) $\frac{12x-1}{8x^2}$

16) $\frac{7y-6}{20}$

17) $\frac{7x-3}{14}$

18) $\frac{3y+14}{y^2-9}$

19) $\frac{x+4y}{6}$

20) $\frac{x+y}{40}$

Lesson 7-9
Page 78

Exercises

1) $\frac{4x}{x^2-1}$

2) $\frac{7x+8}{(x+5)(x-4)}$

3) $\frac{14}{3(x-3)}$

4) $\frac{x+17}{(x-3)(x+1)}$

5) $\frac{7x^2+3x}{(x-3)(x+1)}$

6) $\frac{x^2-6x-15}{(x+5)(x-3)}$

7) $\frac{11x+15}{4x(x+5)}$

8) $\frac{2x+1}{x(x+1)}$

9) $\frac{x(x-1)}{x^2-25}$

10) $\frac{5x-4}{x^2-1}$

11) $\frac{7x+8}{x(x+2)}$

12) $\frac{-3x+8}{x^2-4}$

13) $\frac{5x-8}{(x-4)(x+2)}$

14) $\frac{2x-5}{2x+5}$

15) $\frac{x(3x+11)}{(x+6)(x-1)}$

16) $\frac{3x+20}{x^2-16}$

17) $\frac{5x+5y+8}{(x+y)^2}$

18) $\frac{x(3x-5)}{x^2-1}$

19) $\frac{3x+3y+1}{x^2-y^2}$

20) $\frac{8x+5}{x(x+1)}$

Lesson 7-10
Page 79

Exercises

1) $x=30$

2) $x=12$

3) $x=10$

4) $x=4$

5) $x=12$

6) $x=-18$

7) $x=-7$

8) $x=1\frac{1}{2}$

9) $x=36$

10) $x=21$

11) $x=2$

12) $x=15$

13) $x=8$

14) $x=9$

15) $x=12$

16) $x=24$

17) $x=5$

18) $x=3$

19) $y=16$

20) $x=2$

Solutions

Lesson 8-1
Page 80

Exercises

1) 6
2) 12
3) 16
4) 9x
5) $3\sqrt{2}$
6) $2\sqrt{6}$
7) 7xy
8) $2xy\sqrt{y}$
9) $4\sqrt{3x}$
10) $12xy\sqrt{x}$
11) $2x\sqrt{2}$
12) x^3
13) $6x\sqrt{2}$
14) $6x\sqrt{x}$
15) $xy\sqrt{xy}$
16) $6x\sqrt{y}$
17) $2x^4\sqrt{3}$
18) $5xyz\sqrt{3}$
19) $2xy^2z\sqrt{z}$
20) $2x^2yz^3\sqrt{3y}$

Lesson 8-2
Page 81

Exercises

1) $x = \pm6$
2) $x = \pm6$
3) $x = \pm5$
4) $x = \pm2$
5) $x = \pm4$
6) $x = \pm12$
7) $x = 25$
8) $x = 22$
9) $x = 3$
10) $x = 3$
11) $x = 4$
12) $x = 7$
13) $x = 13\frac{1}{2}$
14) $x = 7$
15) $x = 3$
16) $x = 8$
17) $x = 4$
18) $x = 6$
19) $x = 8$
20) $x = 6$

Lesson 8-3
Page 82

Exercises

1) $11\sqrt{5}$
2) $4\sqrt{5}$
3) $5\sqrt{3}$
4) $7\sqrt{3}$
5) $-3\sqrt{6}$
6) $\sqrt{3}$
7) $25\sqrt{2}$
8) $8\sqrt{2}$
9) $\sqrt{5}$
10) $(2 + 9x)\sqrt{x}$
11) $7\sqrt{3}$
12) $9\sqrt{3}$
13) $4\sqrt{3}$
14) $15\sqrt{5}$
15) $11\sqrt{2}$
16) $-6\sqrt{3}$
17) $8\sqrt{3}$
18) $3\sqrt{5}$
19) 0
20) $3\sqrt{3}$

Lesson 8-4
Page 83

Exercises

1) 4
2) $2\sqrt{6}$
3) $2\sqrt{7}$
4) 90
5) $6\sqrt{6}$
6) $30\sqrt{2}$
7) $10\sqrt{2}$
8) $2\sqrt{7}$
9) 72
10) $5 + \sqrt{55}$
11) 72
12) $5 - 2\sqrt{5}$
13) $36\sqrt{2}$
14) $30\sqrt{3}$
15) 12
16) $3x\sqrt{2}$
17) $33 + 2\sqrt{2}$
18) $6\sqrt{2}$
19) $28 + 10\sqrt{3}$
20) $15 - 6\sqrt{6}$

Lesson 8-5
Page 84

Exercises

1) 3
2) 2
3) $2\sqrt{2}$
4) $\sqrt{5a}$
5) $10\sqrt{3}$
6) $8\sqrt{2}$
7) 3x
8) 27x
9) $42\sqrt{2}$
10) $\frac{x\sqrt{6}}{2}$
11) $4\sqrt{6}$
12) x
13) $\sqrt{3} + \sqrt{5}$
14) 3y
15) 4x
16) 2x
17) $x^2\sqrt{5}$
18) 9xy
19) $2y\sqrt{x}$
20) 8

Lesson 8-6
Page 85

Exercises

1) $\frac{\sqrt{15}}{3}$
2) $6\sqrt{2}$
3) $\frac{\sqrt{2}}{2}$
4) $\frac{5\sqrt{3}}{3}$
5) $4\sqrt{2}$
6) $\frac{5\sqrt{5}}{2}$
7) $\frac{\sqrt{10}}{4}$
8) $\frac{x\sqrt{2}}{2}$
9) 9
10) $\frac{\sqrt{30}}{3}$
11) $\frac{\sqrt{3}}{3}$
12) $\frac{x\sqrt{2}}{4}$
13) $\frac{\sqrt{26}}{2}$
14) $\frac{\sqrt{6}}{2}$
15) $2\sqrt{2}$
16) $\frac{\sqrt{3}}{2}$
17) $3\sqrt{15}$
18) $5\sqrt{2}$
19) $3\sqrt{2}$
20) $\frac{2\sqrt{3}}{3}$

Solutions

Lesson 8-7
Page 86

Exercises

1) $\dfrac{12 - 3\sqrt{5}}{11}$

2) $\dfrac{3\sqrt{10} - 2\sqrt{15}}{3}$

3) $10\sqrt{6} - 20$

4) $4\sqrt{15} + 12$

5) $8 + 2\sqrt{3}$

6) $-9\sqrt{2} + 9\sqrt{7}$

7) $8\sqrt{11} + 8\sqrt{5}$

8) $\dfrac{\sqrt{15} + 7}{-2}$

9) $\dfrac{2\sqrt{13} + 7}{3}$

10) $\dfrac{27 - \sqrt{3}}{22}$

11) $4 - \sqrt{15}$

12) $6 + \sqrt{35}$

13) $\dfrac{-1 + \sqrt{5}}{4}$

14) $3\sqrt{5} + 6$

15) $10 - 5\sqrt{3}$

16) $-3 - \sqrt{7}$

17) $\dfrac{5\sqrt{7} - \sqrt{21}}{22}$

18) $\dfrac{4\sqrt{2} + 2}{7}$

19) $\dfrac{20\sqrt{3} + 4\sqrt{6}}{23}$

20) $20 - 10\sqrt{3}$

Lesson 8-8
Page 87

Exercises

1) $c = 9.90$

2) $c = 13.93$

3) $a = 13.96$

4) $a = 13.23$

5) $a = 6$

6) $c = 34.99$

7) $c = 65$

8) $b = 30$

9) $c = 6.71$

10) $a = 22.91$

11) $a = 21$

12) $c = 25$

13) $c = 6.32$

14) $b = 12$

15) $a = 9.80$

16) $c = 8.60$

17) $c = 17$

18) $b = 8$

19) $a = 8$

20) $c = 39.05$

Lesson 8-9
Page 88

Exercises

1) 6.7

2) 10

3) 10

4) 6.7

5) 5

6) 9.8

7) 6.4

8) 8.5

9) 5

10) 13

11) 10

12) 13

13) 5

14) 8.5

15) 5.4

16) 15

17) 5

18) 13

19) 10

20) 26

Lesson 8-10
Page 89

Exercises

1) $(2, 1)$

2) $(7, -5)$

3) $(0, 0)$

4) $(4, 6)$

5) $(-4, -5)$

6) $(1\frac{1}{2}, -6)$

7) $(-1, 1\frac{1}{2})$

8) $(4, -2)$

9) $(-5, 2\frac{1}{2})$

10) $(3, 0)$

11) $(-1, 3)$

12) $(-3\frac{1}{2}, 2)$

13) $(2\frac{1}{2}, 7\frac{1}{2})$

14) $(-3, -5)$

15) $(2, 3)$

16) $(-3, -2\frac{1}{2})$

17) $(2, 1\frac{1}{2})$

18) $(7\frac{1}{2}, -4)$

19) $(-1, -2)$

20) $(-1, -2)$

Solutions

Lesson 9-1
Page 90

Exercises
1) x = -5 or x = -3
2) x = -1 or x = 5
3) x = -5 or x = 3
4) x = 0 or x = -3
5) x = 0 or x = 8
6) x = -2 or x = 6
7) x = -9 or x = 9
8) x = 0 or x = -8
9) x = -6 or x = 4
10) x = -5 or x = 8
11) x = -8 or x = 9
12) x = 3 or x = 9
13) x = 2 or x = 5
14) x = 0 or x = $1\frac{1}{2}$
15) x = -5 or x = 3
16) x = 4 or x = 7
17) x = 0 or x = -3
18) x = -5 or x = 7
19) x = 2 or x = 6
20) x = -9 or x = -1

Lesson 9-2
Page 91

Exercises
1) x = ±12
2) x = ±6
3) x = ±7
4) x = ±1
5) x = ±8
6) x = ±5
7) x = ±5
8) x = 10 or x = -4
9) x = $3\sqrt{3}$ or x = $-3\sqrt{3}$
10) x = $4\sqrt{2}$ or x = $-4\sqrt{2}$
11) x = $5\sqrt{2}$ or x = $-5\sqrt{2}$
12) x = $\sqrt{7}$ or x = $-\sqrt{7}$
13) x = 3 or x = -1
14) x = $-5 + \sqrt{23}$ or x = $-5 - \sqrt{23}$
15) x = 3 or x = -3
16) x = $\sqrt{2}$ or x = $-\sqrt{2}$
17) x = $5\sqrt{3}$ or x = $-5\sqrt{3}$
18) x = $2\sqrt{3}$ or x = $-2\sqrt{3}$
19) x = $2 + \sqrt{3}$ or x = $2 - \sqrt{3}$
20) x = 2 or x = -1

Lesson 9-3
Page 92

Exercises
1) x = 2 or x = -4
2) x = 7 or x = -3
3) x = 1 or x = 3
4) x = 10 or x = -4
5) x = $2 + \sqrt{11}$ or x = $2 - \sqrt{11}$
6) x = -5 or x = -3
7) x = -15 or x = 1
8) x = $2 + \sqrt{15}$ or x = $2 - \sqrt{15}$
9) x = 4 or x = 2
10) x = $2 + \sqrt{2}$ or x = $2 - \sqrt{2}$
11) x = $2 + \sqrt{5}$ or x = $2 - \sqrt{5}$
12) x = -4 or x = 7
13) x = -2 or x = 6
14) x = -5 or x = 2
15) x = -3 or x = 7
16) x = -3 or x = -1
17) x = 1 or x = 7
18) x = $3 + \sqrt{5}$ or x = $3 - \sqrt{5}$
19) x = $-1 + \sqrt{2}$ or x = $-1 - \sqrt{2}$
20) x = $3 + \sqrt{29}$ or x = $3 - \sqrt{29}$

Lesson 9-4
Page 93

Exercises
1) x = 3 or x = 2
2) x = $\frac{1}{3}$ or x = 2
3) x = $\frac{5 + \sqrt{21}}{2}$ or x = $\frac{5 - \sqrt{21}}{2}$
4) x = 4 or x = -5
5) x = $\frac{1 + \sqrt{21}}{2}$ or x = $\frac{1 - \sqrt{21}}{2}$
6) x = $\frac{1 + \sqrt{33}}{4}$ or x = $\frac{1 - \sqrt{33}}{4}$
7) x = $-\frac{4}{3}$ or x = 2
8) x = $5 + \sqrt{3}$ or $5 - \sqrt{3}$
9) x = -1 or x = 5
10) x = 2 or x = 5
11) x = -1 or x = $-\frac{1}{3}$
12) x = 1 or x = $\frac{3}{5}$
13) x = -1 or x = 3
14) x = $1 + \sqrt{6}$ or x = $1 - \sqrt{6}$
15) x = $-1 + \sqrt{2}$ or x = $-1 - \sqrt{2}$
16) x = $1 + \sqrt{3}$ or x = $1 - \sqrt{3}$
17) x = -3 or x = 7
18) x = $\frac{5 + \sqrt{33}}{4}$ or x = $\frac{5 - \sqrt{33}}{4}$
19) x = $\frac{3 + \sqrt{29}}{2}$ or x = $\frac{3 - \sqrt{29}}{2}$
20) x = $\frac{3 + \sqrt{41}}{4}$ or x = $\frac{3 - \sqrt{41}}{4}$

Solutions

Lesson 9-5
Page 94

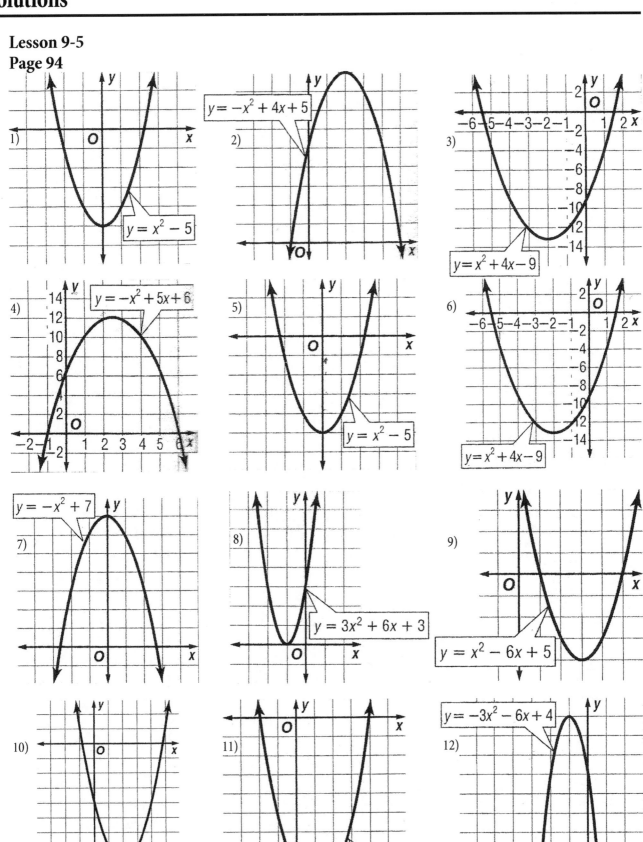

1) $y = x^2 - 5$

2) $y = -x^2 + 4x + 5$

3) $y = x^2 + 4x - 9$

4) $y = -x^2 + 5x + 6$

5) $y = x^2 - 5$

6) $y = x^2 + 4x - 9$

7) $y = -x^2 + 7$

8) $y = 3x^2 + 6x + 3$

9) $y = x^2 - 6x + 5$

10) $y = x^2 - 4x - 4$

11) $y = x^2 - 2x - 8$

12) $y = -3x^2 - 6x + 4$

Solutions

Lesson 9-5, continued

13)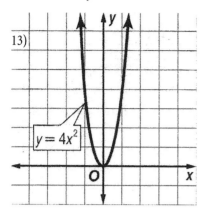

$y = 4x^2$

14)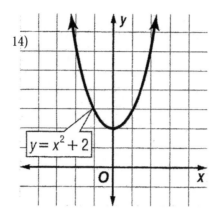

$y = x^2 + 2$

15)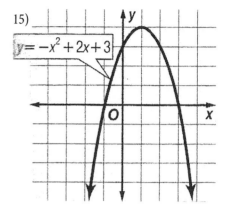

$y = -x^2 + 2x + 3$

16)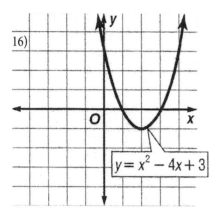

$y = x^2 - 4x + 3$

17)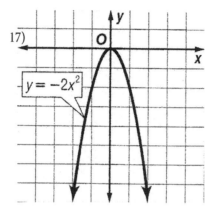

$y = -2x^2$

18)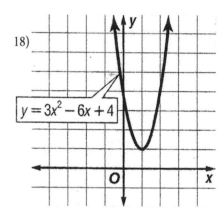

$y = 3x^2 - 6x + 4$

19)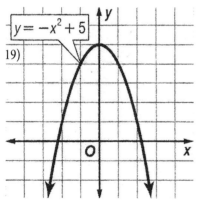

$y = -x^2 + 5$

20)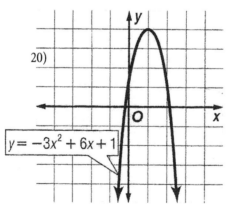

$y = -3x^2 + 6x + 1$

Lesson 9-6
Page 95

Exercises
1) 2
2) 1
3) 2
4) 0
5) 2
6) 0
7) 1
8) 2
9) 1
10) 0
11) 2
12) 0
13) 2
14) 0
15) 1
16) 0
17) 0
18) 0
19) 2
20) 2

Solutions

Lesson 10-1
Page 96

Exercises

1) $3x + 12 = 50$
2) $2x - 40 = 90$
3) $x + 15 = 12$
4) $15 = 4x - 10$
5) $5x - 6 = 2x + 8$
6) $2(x + 6) = 18$
7) $5x + 12 = 95$
8) $4x - 7 = -20$
9) $2x + 20 = 48$
10) $3x + 4 = 5x - 7$
11) $3x = 63$
12) $\frac{x}{6} = 5$
13) $36 = x - 6$
14) $5x + 15 = 7x - 3$
15) $2(x + 5) = 30$
16) $2x + 6 = x - 30$
17) $5x - 7 = 34$
18) $\frac{1}{3}x + 12 = 42$
19) $5x - 7 = 9$
20) $6(x - 8) = 15$

Lesson 10-2
Page 97

Exercises

1) 15
2) 5
3) 8
4) 60
5) 40
6) John = 15, Mike = 3
7) 12 and 24
8) 24, 120
9) 560 boys, 325 girls
10) 125 and 126
11) 89, 90, 91
12) 5
13) 124, 126, 128
14) 9
15) 10, 60

Lesson 10-3
Page 98

Exercises

1) 4 hours
2) $6\frac{2}{3}$ hours = 6 hours 40 minutes
3) 6 hours
4) 8 miles
5) 5 hours
6) 5 hours
7) 4 hours
8) 6 hours
9) 3 hours
10) $2\frac{2}{3}$ hours

Lesson 10-4
Page 99

Exercises

1) 25 lbs. of mixed nuts, 5 lbs. of mixed fruit
2) 25 lbs. of cashews, 20 lbs. walnuts
3) 72 lbs. of chocolate, 48 lbs. of vanilla
4) 27 lbs. peanut butter, 18 lbs. pecan
5) 12 lbs. peanuts, 8 lbs. of cashews
6) 60 quarts of 24% cream, 30 quarts of 18% cream
7) 20 liters
8) 120 liters of water
9) 5 liters of water
10) 42 liters of 30% solution, 28 liters of the 80% solution

Lesson 10-5
Page 100

Exercises

1) $3\frac{3}{5}$ hrs. = 3 hrs. 36 min.
2) $1\frac{1}{5}$ hrs. = 1 hr. 12 min.
3) $1\frac{5}{7}$ hrs.
4) $3\frac{1}{3}$ hrs. = 3 hrs. 20 min.
5) 24 hrs.
6) 6 hrs.
7) 20 hrs.
8) 6 hrs.
9) 8 hrs.
10) $15\frac{3}{5}$ hrs. = 15 hrs. 36 min.

Lesson 10-6
Page 101

Exercises

1) John is 54, Tom is 18
2) Father is 60, Son is 20
3) Jim is 32, Roger is 8
4) Jose is 46, Maria is 26
5) Brian is 40, Lee is 25
6) Joe is 30, Cal is 10
7) William is 24, Father is 48
8) Julie is 40, Kim is 15
9) Mother is 24, Daughter is 6
10) Elena is 18, Katya is 12

Solutions

Lesson 10-7
Page 102

Exercises

1) 4 nickels, 12 dimes

2) 5 pennies, 10 dimes, 15 nickels

3) 3 nickels, 9 dimes, 18 quarters

4) 3 dimes, 7 quarters

5) 70 dimes, 33 quarters

6) 10 nickels, 3 quarters

7) 8 nickels, 40 dimes

8) 7 dimes, 19 nickels

9) 3 nickels, 9 quarters, 8 dimes

10) 12 dimes, 3 nickels

Lesson 10-8
Page 103

Exercises

1) $1200 at 6% interest

2) $4000 invested at 6%, $4500 invested at 7%

3) $4250 at 6%, $8500 at 5%

4) $2000 at 6%, $2250 at 8%

5) $5125 at 6%, $3125 at 10%

6) $1000 at 8.5%

7) $900 at 6%, $300 at 8%

8) $1500 at 6%, $2000 at 8%

9) $1500 at 6%, 500 at 8%

10) $210,000 interest

Solutions

Final Review
Page 104

1) 3
2) 90
3) -18
4) 4
5) $-\frac{5}{12}$
6) $-\frac{5}{8}$
7) $-1\frac{7}{8}$
8) 19
9) 2.48
10) 8.4
11) -46.92
12) -70
13) 4,096
14) 3
15) 1,024
16) 39
17) 17
18) 65
19) 5
20) 64

Final Review
Page 105

21) $2^2 \times 3^2 \times 5$
22) 75
23) 360
24) 4.25×10^{11}
25) 9.6×10^{-5}
26) $x = 1\frac{2}{3}$
27) 585
28) 160
29) 25%

Final Review
Page 106

30) $x = 58$
31) $x = -85$
32) $y = 36$
33) $x = 51$
34) $n = 15$
35) $x = 2$
36) $x = \frac{1}{2}$
37) $n = -1$
38) $m = 2$
39) $x = 7$ or $x = 1$
40) $x = 5$ or $x = -5$
41) $x = 1\frac{5}{6}$
42) $n = -4$
43) $x = -40$
44) $x = -3$

Solutions

Final Review
Page 107

45)

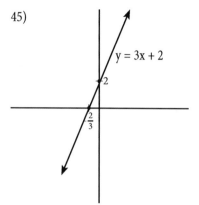

$y = 3x + 2$

46)

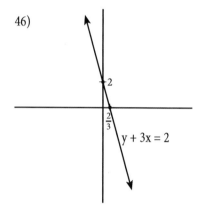

$y + 3x = 2$

47) x-intercept = 3, y-intercept = 4

48) 4

49) $\frac{1}{2}$

50) slope = -3

51) $y = 4x - 5$

52) $y = -2x + 6$

53) $y = 3x + 1$

Final Review
Page 108

54) $y - 6 = 5(x - 4)$

55) $3x + 2y = 6$

56)

57)

58)

59)

60)

61) $x \le 2$

62) $x \le -55$

63) $x > 56$

64) $-7 \le x \le -1$

65) $x < -12$ or $x > 2$

66) $x < -8$ or $x > 8$

67) $-5 < x < 5$

Final Review
Page 109

68)

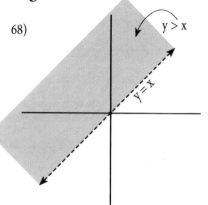

$y > x$

$y = x$

69)

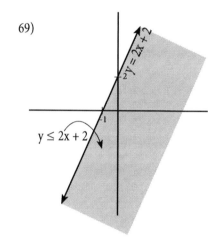

$y = 2x + 2$

$y \le 2x + 2$

70) (2, 10)

71) (1, -1)

72) (-2, 3)

73) (6, 4)

74)

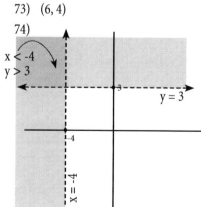

$x < -4$
$y > 3$

$y = 3$

$x = -4$

75) $-12x^2 - 8xy - 3y^2$

76) $-4x^2 - 2x + 3$

77) $7x - 4$

Solutions

78) $4x^2 - 12xy - 10y^2$

79) $-28x^4y^7$

80) $63x^5y^6$

81) $-9xy^3z^6$

82) $3x^3y^2z^2$

83) $-6x^4 + 14x^3 - 20x^2$

84) $-10y^3 - 35y^2$

85) $n^2 + 13n + 42$

86) $36x^2 + 51x + 18$

87) $2x^2 + x - \dfrac{5}{2x}$

88) $3x^3 + x^2 - 4x$

89) $15xy(x - 2)$

90) $x(x + 3)(x + 8)$

91) $(x - 9)(x + 5)$

92) $4(5x + 3y)(5x - 3y)$

93) $7(x + y)(x - y)$

94) $50(x + 4)(x - 2)$

95) $\dfrac{7}{8}$

96) $\dfrac{(x - 5)}{(x + 5)}$

97) $\dfrac{3}{x + 5}$

98) $\dfrac{x - 1}{x - 8}$

99) $\dfrac{1}{x + 2}$

100) $\dfrac{x - 2}{x + 2}$

101) $x = -1$

102) $\dfrac{x}{x + 3}$

103) $x = 5$

104) $\dfrac{x}{2y^2}$

105) $\dfrac{3(x + 4)}{4(2x - 9)}$

106) $\dfrac{8x - 1}{2x + 1}$

107) $\dfrac{2x + 5}{x - 2}$

108) $\dfrac{3 + 5x}{x^2}$

109) $\dfrac{4}{3}$

110) $\dfrac{15}{4x}$

111) $\dfrac{5x + 21}{6}$

112) $\dfrac{2y}{15}$

113) $\dfrac{44 - x}{15}$

114) $x = 8$

Solutions

Final Review
Page 113

115) $x = 6\frac{2}{3}$

116) $x = 3\frac{3}{4}$

117) $5\sqrt{3}$

118) $3x\sqrt{2x}$

119) $4x\sqrt{5}$

120) $12x^2y$

121) $\frac{7\sqrt{5}}{5}$

122) 3

123) $\sqrt{5}$

124) $13\sqrt{5}$

125) $6\sqrt{2}$

126) $3\sqrt{10}$

127) 8

128) $5xy\sqrt{2}$

Final Review
Page 114

129) $x = 8$ or $x = -8$

130) $x = 13$ or $x = -13$

131) $x = 9$ or $x = -9$

132) $\frac{12 + 2\sqrt{3}}{33}$

133) $\frac{15 + 3\sqrt{2}}{23}$

134) $c = 25$

135) $b = 18$

136) $D = 17$

137) $M = (12, 5)$

Final Review
Page 115

138) $x = 9$ or $x = -9$

139) $x = 10$ or $x = -10$

140) $x = -4$ or $x = 7$

141) $x = 6$ or $x = -1$

142) $x = -1$ or $x = 4$

143) $x = 2$ or $x = -10$

144) $x = 13$ or $x = -1$

145) $x = 7$ or $x = -2$

146) $x = 1 + \frac{\sqrt{6}}{2}$ or $x = 1 - \frac{\sqrt{6}}{2}$

147) $x = 5 + \sqrt{3}$ or $x = 5 - \sqrt{3}$

148) $x = 2 + \sqrt{3}$ or $x = 2 - \sqrt{3}$

Final Review
Page 116

149)

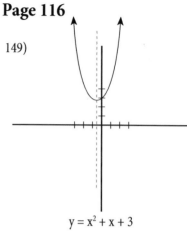

$y = x^2 + x + 3$

150)

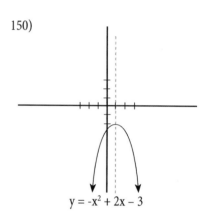

$y = -x^2 + 2x - 3$

151) 6

152) Ron 84 km, Rhonda 88 km

153) Pecans 40 lbs., pistachios 80 lbs.

154) $1\frac{1}{5}$ hrs. = 1 hr. 12 min.

155) $6,000 at 5%, $14,000 at 7%

156) Mother is 35, daughter is 7

157) 7 nickels, 28 dimes, 4 quarters

$-2t(8t-21)$

$2.62 = t$

$\dfrac{-7 \pm \sqrt{(7)^2 - 4(-1)(4)}}{2(-1)}$

$\dfrac{42}{32}$

$-16t^2 + 42t$

$-7 \pm \sqrt{49 + 16}$

$-16t^2 + 4$

7^2

$-12 \pm \sqrt{4}$

$49 - 4(2)(3) = 49 - 29 = 20$

$-12 \pm$

$\dfrac{-12 \pm \sqrt{144 - 48}}{8}$

$\dfrac{-4 \pm \sqrt{(4)^2 - 4(4)}}{2(4)}$

$\dfrac{-12 \pm 9.8}{8}$

$\dfrac{-12 \pm \sqrt{12^2 - 4(4) \cdot 3}}{2(4)}$

$\dfrac{13}{2}$

$-16t(t - 13)$

$-16t^2 + 208t + 0$

$$f(x) = 2(x^2 - 8x) + 10(-4)^2$$

$$2(x^2 - 8x + (-4)^2) + 17 - 15$$

$$2(x^2 - 8x + 16) - 15$$

$$2(x - 4)^2 - 15 = 0$$

$$2(x - 4)^2 = 15$$

$$(x - 4)^2 = 7.5$$

$$4 \pm \sqrt{7.5} = x$$

$$\frac{-28}{(2)-1} \quad 14$$

$$-(x^2 - 20x + 100) - 75 + 100$$

$$-(x + 10)^2 + 25 - 25$$

$$x = 10 \pm 5$$

$$25$$

$$-(x^2 + 28x)$$

$$-(x^2 - 28x) - 192$$

$$-(x^2 - 28x + 14^2) - 192 + 14^2$$

$$-(x + 14)^2 + 4 = 0$$

$$(x + 14)^2 = 4 \qquad x + 14 =$$

$$x = 14 \pm 2$$

$$\sqrt{4} = 2$$

$$-(x^2 - 60 + 30^2) \, \overline{\Theta / 16} + 30^2$$

$$-(x + 30)^2 \, \underset{784}{\overset{900}{}} = 0 \qquad \frac{21.04}{4} = 5.26$$

$$(x+30) \qquad = 28 \qquad \frac{9 \pm 12.04}{4}$$

$$\frac{-7 \pm \sqrt{(7)^2 - 4(-1)(9)}}{2(-1)} X = 30 \pm$$

$$36 -$$

$$1 \pm \sqrt{(-1)^2 - 4(-12)}$$

$$X = \frac{b \pm \sqrt{b^2 - 4ac}}{2a} \qquad 25x^2 = -36 \qquad 9 \sqrt{81 + 64}$$

$$2x^2 + 9x - 8$$

$$X = \frac{-(-1) \pm \sqrt{(-1)^2 - 4(2)(-6)}}{2(2)} \qquad \frac{9 \pm \sqrt{(9)^2 - 4(2)(-8)}}{\dfrac{9 \pm \sqrt{81 - 8(-8)}}{4} \quad 2(2)}$$

$$3b$$

$$1,2,3,4,6$$

$$\frac{4x^2 + 6x + 6x + 9}{2x^2} \quad \frac{3}{24}$$

$$\frac{2(x^3 + 7x^2)}{x^2} \frac{(2x + 14)}{2} \qquad 4/2$$

$$(x + 7)(x^2 + 2) \qquad 3x^2 + 10x$$

$$3x(3x^2 + 6x - 6) \qquad 3x$$

$$3x(x^2 + x - 6) \frac{(x^2 - 2x)}{x} \frac{(-4x + 8)}{-4}$$

$$\frac{(3x^2 + 6x + 4x + 8)}{3x} \quad (x - 2)$$

$$x + 2 \qquad \frac{(x^2 - 8x)}{(x) - 8} \frac{(3x - 24)}{(3)x - 8}$$

$(ax + ab) \cdot (2ay + 2xy)$ $\qquad$ $-32 = (x+9)^2$

$3(x^2 + 18x + 9^2) + 15 + 81$ $\qquad$ 32

$-96 = 3(x + 81)^2$ $\qquad$ $96 - 96 = 3(x+3)^2$

$-96 = 3(x+9)^2$ $\qquad$ 1

-14

$-(x^2 + 30x) - 125$ $\qquad$ -32

$-(x^2 + 30x + 15^2) - 125 + 225$

$f(x) = -(x+15)^2 + 100$

$15 \pm 10 = x$

$0 = -(x+15)^2 + 100$

-100

$-100 = -(x+15)^2$ $\qquad$ $10 = -(x+15)$

$$(-3x^2 + 18)(x + 5)$$

$$\frac{(x^3 - x^2)}{x^2} - \frac{(16x + 16)}{16}$$

$$(x - 1) \cdot (x - 1)$$

$$(x^2 + 16)(x - 1)$$

$$(3x^2 + 12)(x + 12)$$

$$-3(x +$$

$$-3(x^3 + 5x^2 - 4x - 20)$$

$$-3(x^3 + 5x^2)(-4x - 20)$$

$$-3(x + 5)$$

Made in the USA
Lexington, KY
19 December 2017